BOEING **HORNET** SQUADRONS

BOEING HORNET SQUADRONS

THE LEGACY SERIES

PETER FOSTER

First published 2019

The History Press
The Mill, Brimscombe Port
Stroud, Gloucestershire, GL5 2QG
www.thehistorypress.co.uk

British Library Cataloguing in Publication Data.
A catalogue record for this book is available from the British Library.

ISBN 978 0 7509 8558 1

Typesetting and origination by The History Press
Printed and bound in India by Thomson Press India Ltd

Above: F/A-18C 165186 diving into Rainbow Canyon, California, in October 2014.

CONTENTS

Preface 6
Introduction 7
Acknowledgements 14

1 United States Navy Fleet Squadrons 15
2 Naval Strike and Air Warfare Centre (NSAWC) 69
3 Blue Angels 71
4 United States Marine Corps Squadrons 73
5 Test and Evaluation Squadrons 101
6 NASA 105
7 Carrier Air Wings 107

Production List 109
United States Aircraft Carriers 115
Abbreviations 118
Sources 119

PREFACE

Back in November 1988 I was privileged to be given the opportunity to fly in a Hornet aircraft operated by the Canadian Forces from Baden-Soellingen air base in Germany. My pilot for the day was the No. 4 Wing Commander, Colonel Jean Boyle, who later went on to become the Canadian Chief of Defence Staff.

This was the most futuristic fighter aircraft I had flown in at that time and, apart from the superb flight, fantastic weather conditions and awesome photographic opportunities, it began my love affair with this aircraft.

Since then I have photographed numerous Hornets from all of the operating countries both in the air-to-air environment and on the ground. This has created a great personal portfolio of this impressive aeroplane that prompted my desire to publish this photographic tribute.

Originally the concept was to produce a single volume of all the variants and all the operators but I soon came to realise that it would be just too vast for one book. I then decided to split between the legacy series and the Super Hornets but this too created a space problem to do the subject justice. I have therefore restricted the first book to just the United States' use of the legacy series of aircraft and hopefully in due time I will be in a position to follow up with one on the export customers and then finally the Super Hornets.

The use of the legacy series by United States forces is very much on the decline and the few remaining operators will either soon see a transition the futuristic Lockheed F-35 or pass into reserve status because of well-publicised age-related serviceability issues.

The aircraft is an impressive performer at any level and I hope the images contained in this book will show this adequately while providing a history of its use with the US forces. I must thank all the service personnel who have directly or indirectly assisted in providing these photographic opportunities and to a number of friends who have allowed me to use one or two of their images to complete the story where my own portfolio was lacking.

I hope the reader will find this a good book for reference and enjoyment.

Peter R. Foster
Doddington, UK

INTRODUCTION

The F-18 Hornet is very much a Boeing, or should I say McDonnell Douglas, success story because it is not only a 'jack of all trades' but very much a master of two in that it is both a very capable fighter, integrating a wide range of systems, and a superb bomber capable of delivering a large and varied load of ordnance. Its two-engine reliability and rugged construction has made it an ideal strike fighter for both carrier-based operation and forward base deployment, something that the aircraft it replaced perhaps lacked.

The Hornet owes its origins to yet another manufacturer, Northrop, who designed and produced the YF-17, a contender for the USAF's Lightweight Fighter (LWF) programme, the outcome of which saw its main opponent, the General Dynamics YF-16, crowned the winner. However, the Navy and US Marine Corps were at the time looking towards replacing a number of front-line types, namely the F-4S Phantom II, A-7E Corsair II, A-6E Intruder and A-4M Skyhawk, and saw the YF-17 concept as a good possibility.

The YF-17, unburdened by the systems and ordnance of a front-line fighter, was like a race car over a family saloon. That said, its potential was well recognised by those who had the opportunity to fly it. The sleek shape and a high performer, its main Achilles heel was that it was powered by an unproven engine, the GE YJ101 (later F101), and it had two of them. From a cost perspective this was perhaps to be the overriding reason that the LWF was awarded to the F-16, which with one more tried and tested engine was seen as a safer and less costly solution.

The US Navy was at the time fully committed to the Grumman F-14 Tomcat as its main front-line fighter and some saw the YF-17, or F-18 as it eventually became, as a threat to that programme. However, akin to the United States Air Force's (USAF) own issues over its purchase of the F-15 Eagle, the Navy was instructed to consider the LWF fighters as an alternative to the high end cost of the F-14.

In monetary terms the YF-17 was seen as a cheap alternative and the US Navy eventually bowed to the inevitable of having to procure such a fighter to complement and work alongside the much bigger and more expensive Tomcat. However, Northrop itself had little experience in designing aircraft for carrier-based operations so the company decided to accept an offer from McDonnell Douglas to collaborate on the YF-17's 'navalisation'.

The marriage between these two heavyweight aircraft manufacturers was not always bliss. The Northrop designers, having a mass of expertise on lightweight designs, were to some degree aghast as the YF-17 began to gain weight and drag as it was redesigned for carrier operations.

The basic shape was retained but the fuselage of the emerging F-18 Hornet saw an increase in girth to allow for additional internal fuel to meet the range requirements of a naval fighter. The nose was redesigned to accommodate the 28in radar dish required for enhanced coverage and the jet was fast becoming a much better equipped war fighter over the sleek demonstrator. The unproven F101 engines were themselves providing better performance and reliability levels, with the redesign now marketed as the F404-GE-400 power plant providing some 16,000lb (7,257kg) of thrust.

F/A-18A+ 162872 of VFA-87 seen just before dark on 'Sierra' air-to-air refueling track over Afghanistan. The squadron swapped its C model jets for the upgraded A just before this deployment. Later it was to exchange jets with VFC-12.

VFA-125's executive officer-assigned aircraft seen at NAS Lemoore in October 2009 just prior to its retirement to AMARG. 163443 saw previous service with VFA-86 as the CO's aircraft in September 1991, when it visited the UK with the carrier air wing.

It was at this point that the Navy made plans for a buy of some 780 aircraft split between the conceived F-18 fighter and a A-18 bomber version, and an initial order was placed for nine single seat F-18A and two twin-seat TF-18A prototypes. The latter were to later be re-designated F-18B.

Initial testing and trials were carried out under a principle site concept rather than spread around different facilities, with both navy and civilian personnel working alongside one another and providing a much more efficient flow of information. This was undertaken at the Naval Air Test Centre (NATC) at Patuxent River, Maryland.

There were still many hurdles to overcome, none more crucial than the first carrier landing. No end of simulations can replicate this and on 30 October 1979 the jet was flown aboard the USS *America* (CV-66) for the first time. The trial lasted five days and saw two Navy test pilots undertake seventeen touch-and-go landings and thirty-two arrested landings and take-offs involving some fourteen hours of flying without a single hitch.

Into Service

As with most new programmes, there were a number of issues that had to be addressed as trials progressed but the biggest was cost as the price was steadily increasing to a point where it was nearing that of the more sophisticated F-14A Tomcat, negating the original concept of a cheap stablemate. Relations between the two manufacturers were also becoming strained, resulting in a number of law suits being issued, although to some degree this became just a side issue to the ongoing development.

The requirement for the jet had risen to some 1,366 aircraft with the US Marine Corps now on side. Delivery of the first aircraft for the Navy took place in May 1980 when the Fleet Replacement Squadron (FRS) VFA-125 Rough Riders at NAS Lemoore, took delivery of its first Hornet. The aircraft, 161250, a TF-18A at that time, wore the code 'NJ' and modex '500' with Navy titles on one side and Marine on the other.

As deliveries increased, so came the need for more training capacity. First came an east coast FRS, VFA-106 Gladiators at NAS Cecil Field in Florida, then the USMC stood up its own training unit in the shape of VMFAT-101 at El Toro in California. Ironically, the very first non-training squadron to begin re-equipping with the Hornet was a USMC one, VMFA-314 'Black Knights'.

This took place in January 1983 at MCAS El Toro and the Navy was to follow suit ten months later when VFA-113 Stingers at NAS Lemoore took delivery of its first jet.

The US Navy was initially to transition both LTV A-7E Corsair II and Grumman A-6E Intruder units onto the now designated F/A-18A Hornet, leaving the former F-4S Phantom II units to convert to the highly sophisticated Grumman F-14A Tomcat. The exceptions were VF-151 and VF-161, which became the only two front-line fleet USN Phantom users to move over to the Hornet. The USMC, on the other hand, was to transition all of its units over onto a single type, with the exception of its Grumman EA-6B Prowler squadrons, one of which, VMAQ-2, continued to operate this suppression of enemy defences (SEAD) aircraft until completing its final operational detachment on 17 November 2018. The squadron was scheduled to stand down in early 2019.

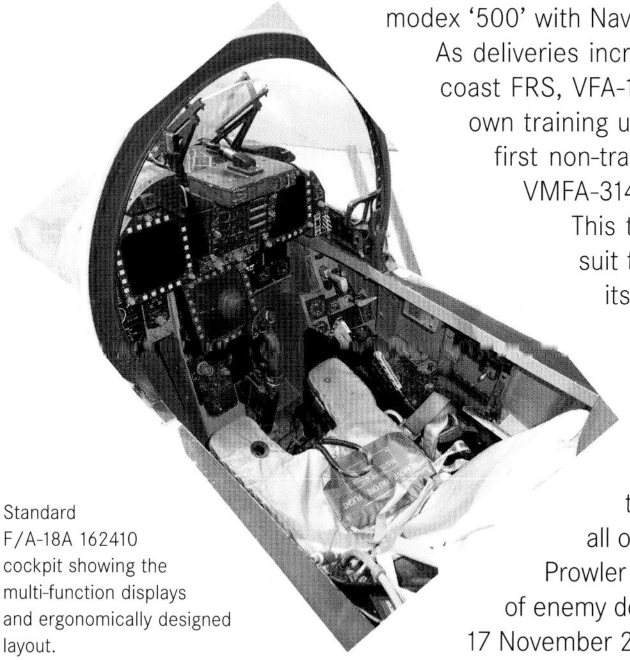

Standard F/A-18A 162410 cockpit showing the multi-function displays and ergonomically designed layout.

F/A-18C 165186 belonging to VMFA-232 low level in 'Star Wars Canyon', California, during a normal training sortie during October 2014. This is a block 48 aircraft from Lot 18, one of the last C models to be procured.

In the end the US Navy/US Marine Corps were to procure some 380 and 40 first generation Hornets of the F/A-18A and B designation respectively to re-equip fifty-nine squadrons of both the front-line and reserve components. Later improvements in flight instrumentation and avionics saw from block 21 the introduction of the improved F/A-18C/D version. The delivery of the first F/A-18C formally took place on 21 September 1986 when the aircraft was delivered to NATC (later re-designated VX-23) at Patuxent River. A second jet was delivered to the Naval Air Weapons Centre at China Lake two days later. VFA-113 Stingers also became the first fleet squadron, alongside VFA-25 'Fist of the Fleet', to adopt the C version in the final quarter of 1989.

The improved C/D versions were also equipped to carry additional types of ordnance over the earlier A/B versions. These missiles, the AIM-120 AMRAAM, AGM-84 Harpoon and AGM-65F infrared imaging Maverick, gave the force added teeth along with the ability to deploy the Hughes AN/AAR-50 thermal navigation pod, the Loral AN/AAS-38 NITE Hawk FLIR (forward-looking infrared array) targeting pod, night vision goggles, and two full-colour (formerly monochrome) multi-function display (MFDs) with a colour moving map. As time moved on some earlier versions of this 'legacy' Hornet were to receive a similar upgrade as F/A-18A+ jets, although the C/D model became the front-line choice.

From 1993, the AAS-38A NITE Hawk added a designator/ranger laser, allowing it to self-mark targets. The later AAS-38B added the ability to strike targets designated by lasers from other aircraft.

The US Navy only used the single-seat A and C models for the fleet units, while the USMC saw far more potential for the two-seat B and D, particularly in the night-attack concept. However, some F/A-18A+ and F/A-18C aircraft benefitted from the introduction of forward-looking infrared (FLIR), night vision goggles (NVG) and a colour digital map display, although such improvements did not see any designation change.

The first night-attack F/A-18D, 163986, was handed over to NATC at Patuxent River on 1 November 1989 in partnership with the first F/A-18C similarly equipped, 163985. The first operational night-attack squadron was VMA(AW)-121 Green Knights, which was re-designated VMFA(AW)-121 and took place to some extent in a controversial equipment change when the USMC opted to give up its entire fleet of Grumman A-6E Intruder aircraft in favour of the F/A-18D. The USMC was to eventually equip twenty-four squadrons with the legacy Hornet.

The sixty D-model Hornets are now configured as the night-attack F/A-18D (RC) with ability for reconnaissance. These could be outfitted with the ATARS electro-optical sensor package that includes a sensor pod and equipment mounted in the place of the M61 cannon.

Seen high over the Atlantic seaboard of the United States during the USS *George Washington* shakedown cruise in March 1996, this jet, 165213 'AG-205' from VFA -131, was delivered factory fresh to the unit. However 131 has since transitioned to the Super Hornet and has passed its mounts on to VFA-34.

30 Years On

It seems incredible that the McDonnell Douglas/Boeing F/A-18 Hornet has been in front-line service for more than thirty years, having supplanted not only its intended types of F-4, A-4, A-6 and A-7 but also the F-14 Tomcat, the stablemate it was intended to complement. True we are, in the case of the US Navy, seeing this occur through the introduction of the bigger and more capable Super Hornet but there are still a few Navy units operating the older legacy version and these will probably continue to do so until replaced by the fifth generation Lockheed F-35C Lightning II.

Having served through two major wars in the Gulf and countless smaller skirmishes, these legacy aircraft are becoming very tired. This has seen a number of short-term fixes. Initially, some older F/A-18A airframes with lower hours and fewer traps were given avionics upgrades consistent with the newer C models and more latterly through F-35 development delays there has been an increased purchase of the F/A-18E/F Super Hornet.

A report in *Sea Power* magazine also stated: 'The Navy has set in motion plans to shut down the staff of Carrier Air Wing Ten (CVW-10) and three carrier-based aviation squadrons. The moves come even as political momentum builds to grow the size of the US fleet, including building the aircraft carrier fleet back to 12 carriers.'

The Navy deactivated Carrier Air Wing Fourteen (CVW-14) effective from 31 March 2017, even though such initial plans had earlier been reversed.

The 'Night-attack' F/A-18D version of the Hornet was seen as a priority for the US Marine Corps. Late production aircraft 165411, from Lot 20, is a block 51 airframe and serves with VMFA(AW)-225 Vikings from MCAS Miramar, from where it is seen taking off.

The wing, based at Naval Air Station (NAS) Lemoore, California, had been under-performing in squadron strength and had not deployed since 2011. The 2017 National Defense Authorization Act allowed the Navy to reduce the number of carrier air wings to nine until the number of deployable carriers could fully support a tenth wing, or until 1 October 2025, whichever scenario were to arrive first.

The three squadrons deactivated as a result included Strike Fighter 15 (VFA-15), an F/A-18C squadron based at NAS Oceana in Virginia Beach, Virginia. VFA-15 was deactivated on 31 May, although it is understood that the squadron gave up its last aircraft in December 2016, with its the last flight occurring on the 22nd of that month.

A report in the 15 November issue of the *Navy Times* confirms that the US Navy had conducted their last combat deployment of the 'Legacy Series' Hornet aboard the aircraft carrier USS Carl Vinson eight months previously, in March 2018. It further stated that the remaining first-generation Hornets will continue to be used by the Reserve units and for aviator training.

For the US Marine Corps the issues have been more difficult, having opted not to go down the Super Hornet route and hang in with its legacy series until the arrival of both the F-35B and F-35C Lightning II. The deactivation of CVW-10 did, however, free up much-needed extra legacy series aircraft.

A report in the April 2017 *Stars and Stripes* stated, 'The Marine Corps will replace its aging fleet of F/A-18 Hornet aircraft with F-35 Lightning IIs more quickly than initially planned because of continued struggles to keep the legacy fighter jets airworthy.'

William E. Taylor, assistant deputy commandant for aviation, told members of the House Armed Services' Readiness subcommittee that at first the Marine Corps intended to field the F-35s to squadrons now flying Hornets and AV-8B Harrier aircraft on similar timetables. This plan changed to instead replacing just its F/A-18s as the priority, transitioning to the F-35 as swiftly as possible.

Given that the Harrier is faring better than the Hornet fleet, three squadrons currently flying Hornets are now planned to transition to F-35s during the next three years. This will result in at least a four-year gap before the next Harrier squadron receives the fifth-generation fighters, though it could be even longer.

Taylor said the Marine Corps was looking into a transition of at least five Hornet squadrons to F-35s before replacing any further Harrier units. There are currently twelve tactical fighter squadrons flying Hornets, five flying Harriers, and three flying F-35s.

The Hornet's many problems have been widely written about. Crashes in 2016–16 killed three airmen and saw the loss of five aircraft, with as many as twelve further incidents following across 2017. Nearly two dozen previously retired Hornets were scheduled to have been pulled from the 'boneyard' (AMARG) at Davis-Monthan Air Force Base in Arizona to be put back into to the fleet, although to date only sixteen are known to have taken that route.

Taylor also confirmed that there would be an assessment of the Hornet fleet, which under the service's present strategy would not be entirely replaced by F-35s until 2030. Hornets, some of which are up to 28 years old, are frequently judged to be too costly to repair, with such severe defects as corrosion typically being exposed by the repairers.

In contrast to this, the Marines' AV-8B Harrier fleet has shown itself to be made of tougher stuff, not least with regard to the ready quantity of spares freely available due to the cut-price purchase of the Royal Air Force's entire Harrier fleet. The USMC AV-8B is scheduled to remain in the fleet through 2031.

The Pentagon's present budget does not permit F-35s to be fielded quickly, a source of frustration for the Marines. Marine Lt Gen. Jon Davis, the deputy commandant for aviation, said that if the process were sped up there could be significant savings in maintenance and operations costs that were being incurred to keep the F/A-18s flying.

More recently, Marine Commandant Gen. Robert B. Neller said: 'In fact, right now, we've got too many Hornets; we've got too many airplanes. We need to get rid of them because we don't have time to fix them.' He commented that often aircraft such as the UH-1Y, Cobra, Harrier and Hornet became redundant as the Corps transitioned to new aircraft platforms.

Clearly the legacy Hornet, production of which ended in 2000, has reached its pinnacle and is now very much in decline. This book will be a testament to all the US users of the variant. For overseas customers operating the F/A-18A to D series, being only land based and not suffering from the rigours of carrier-based operation the issue of replacement is perhaps a little way off, and I hope to be in a position to cover these operations in the future. I also hope to do the same in respect of the more advanced F/A-18E/F Super Hornet and EA-18G Growler derivatives, of which to date some 297 E, 276 F and 125 G models have been delivered with a further thirty plus still on order.

ACKNOWLEDGEMENTS

The majority of the images contained in this book were taken by the author. However, a few have been loaned by other individuals and I must say thanks to Robbie Shaw, Steve Hill, Dan Stijovich, Scott Rathbone, Tony Osborne, Frank Mirande, the late Geoffrey Rhodes, Andy Thomson, Don Jay, Toshiki Kudo and Alec Molton for allowing me access to and the option to use their images.

1

UNITED STATES NAVY FLEET SQUADRONS

VFA-15 Valions

VA-15 transitioned to the F/A-18A Hornet in late 1986, in doing so, re-designating Strike Fighter Squadron Fifteen (VFA-15). The Valions took on their first F/A-18 Hornet in January 1987 and joined Carrier Air Wing Eight (CVW-8) aboard CVN-71.

Some of their earliest first sorties were flown in support of Operation Desert Storm starting in January 1991. One year later, VFA-15 transitioned to the F/A-18C night-attack Hornet. At this point the squadron started developing pioneering tactics such as the use of the Night Vision Goggle (NVG) compatible cockpit and targeting pod. Trials began aboard CVN-74 in 1996 before deploying with the rest of the wing aboard CV-67, where the Valions undertook two periods before once again returning to CVN-71.

The squadron continued combat operations in the spring of 1999 from the flight deck of the USS *Theodore Roosevelt* (CVN-71), where VFA-15 flew more than 300 combat sorties in support of Operation Allied Force. At the end of the cruise the Valions relocated to NAS Oceana.

Operation of the legacy Hornet saw the squadron, according to *Go Navy*, undertake sixteen deployments up to 2014, with a number in support of ongoing operations including missions into Afghanistan in support of Operation Enduring Freedom. In 2001 the squadron flew 185 sorties and 795 hours, releasing 232,000lb of laser-guided bombs (LGB), Joint Direct Attack Munitions (JDAM) and Maverick air-to-ground missiles in just four weeks.

This was followed in 2003 when the squadron delivered more than 245,000lb of ordnance against Iraqi military facilities, air defence sites and terrorist camps. This operation was undertaken in support of Operation Iraqi Freedom. In September 2005, with the rest of the Battle Group (CSG-2), the squadron undertook flight operations in support of the Multi-National Force (MNF) in Iraq. Over the ensuing four months VFA-15 gave vital support to the MNF from Basra to Mosul, and Al-Qa'im to Baghdad. The Valions were the leading squadron within CAG-8, with more than 1,500 sorties and 5,000 flight hours.

Most recently, the Valions deployed aboard the USS *George H.W. Bush* (CVN-77) from May to December 2011 and managed to keep up a 100 per cent combat sortie completion rate while flying 430 missions in support of Operations Enduring Freedom and New Dawn. After two deployments on CVN-77, VFA-15 returned to Oceana in November 2014.

Aircraft assigned to the Valions conducted the squadron's final flight on 22 December 2016 before its scheduled deactivation in March 2017.

Seen at RAF Abingdon in September 1988 shortly after the squadron converted to the Hornet, 163124 transferred to the USMC after VFA-15 upgraded to the C model and now serves with VMFA-115.

VFA-15 has been continuously assigned as part of Carrier Air Wing Eight (CVW-8) and the jets have over the years interchanged with its sister squadron, VFA-87. During the second Gulf War it was tasked with night operations from the *Theodore Roosevelt* and 164673 is seen at the height of the operation.

F/A-18C(N) 164256 AJ-304 from VFA-15 operating as 'Melee 34' over Afghanistan.

VFA-15	25 Aug 1988–11 Oct 1988	F/A-18A	CVN-71	CVW-8	AJ-3xx	NorLant
VFA-15	30 Dec 1988–30 Jun 1989	F/A-18A	CVN-71	CVW-8	AJ-3xx	Med
VFA-15	19 Jan 1990–23 Feb 1990	F/A-18A	CVN-72	CVW-8	AJ-3xx	Shakedown (SoLant)
VFA-15	28 Dec 1990–28 Jun 1991	F/A-18A	CVN-71	CVW-8	AJ-3xx	Desert Shield & Desert Storm
VFA-15	11 Mar 1993–08 Sep 1993	F/A-18C(N)	CVN-71	CVW-8	AJ-3xx	Med
VFA-15	22 Mar 1995–22 Sep 1995	F/A-18C(N)	CVN-71	CVW-8	AJ-3xx	Med
VFA-15	02 Mar 1996–28 Mar 1996	F/A-18C(N)	CVN-74	CVW-8	AJ-3xx	Shakedown(SoLant)
VFA-15	17 May 1996–28 Jun 1996	F/A-18C(N)	CV-67	CVW-8	AJ-3xx	NorLant
VFA-15	29 Apr 1997–28 Oct 1997	F/A-18C(N)	CV-67	CVW-8	AJ-3xx	Med, Persian Gulf
VFA-15	26 Mar 1999–22 Sep 1999	F/A-18C(N)	CVN-71	CVW-8	AJ-3xx	Med, Persian Gulf
VFA-15	25 Apr 2001–10 Nov 2001	F/A-18C(N)	CVN-65	CVW-8	AJ-3xx	Med, Persian Gulf
VFA-15	06 Jan 2003 – 29 May 2003	F/A-18C(N)	CVN-71	CVW-8	AJ-3xx	Caribbean, Med
VFA-15	01Sep2005–11 Mar 2006	F/A-18C(N)	CVN-71	CVW-8	AJ-3xx	Med, Persian Gulf
VFA-15	08 Sep 2008–18 Apr 2009	F/A-18C(N)	CVN-71	CVW-8	AJ-3xx	Med, CENTCOM AOR
VFA-15	11 May 2011–10 Dec 2011	F/A-18C(N)	CVN-77	CVW-8	AJ-3xx	Med, north Arabian Sea, Persian Gulf
VFA-15	15 Feb 2014–15 Nov 2014	F/A-18C(N)	CVN-77	CVW-8	AJ-3xx	Med, north Arabian Sea, Persian Gulf

VFA-22 Fighting Redcocks

VFA-22, the Fighting Redcocks, transitioned from the LTV A-7E Corsair II in 1990. The squadron is now based at NAS Lemoore, California, and operates the two-seat F/A-18F Super Hornet.

Initial transition to the Boeing Hornet came in 1990 as part of CVW-11 aboard CVN-72, USS *Abraham Lincoln*, with the night-attack version of the Hornet fighter, F/A-18C(N), a mount it was to retain for fourteen years predominantly within the structure of Wing 11.

During that period the squadron undertook thirteen deployments at sea with the legacy version of the Hornet. The last, in January 2003, was as part of CVW-9 and the unit then returned to NAS Lemoore to begin transition to the much-improved second generation Hornet that it was eventually to take aboard the USS *Ronald Reagan*, CVN-76.

VFA-22	12 Mar 1991–16 Apr 1991	F/A-18C(N)	CVN-72	CVW-11	NH-3xx	EastPac
VFA-22	28 May 1991–28 Nov 1991	F/A-18C(N)	CVN-72	CVW-11	NH-3xx	WestPac, Persian Gulf
VFA-22	14 Oct 1992–13 Nov 1992	F/A-18C(N)	CVN-72	CVW-11	NH-3xx	EastPac
VFA-22	13 Jan 1993–12 Feb 1993	F/A-18C(N)	CVN-72	CVW-11	NH-3xx	EastPac
VFA-22	02 Mar 1993–28 Mar 1993	F/A-18C(N)	CVN-72	CVW-11	NH-3xx	EastPac
VFA-22	15 Jun 1993–15 Dec 1993	F/A-18C(N)	CVN-72	CVW-11	NH-3xx	WestPac, Persian Gulf, Indian Ocean
VFA-22	11 Apr 1995–09 Oct 1995	F/A-18C(N)	CVN-72	CVW-11	NH-3xx	WestPac, Persian Gulf
VFA-22	11 Oct 1996–11 Apr 1997	F/A-18C(N)	CV-63	CVW-11	NH-3xx	WestPac, Persian Gulf
VFA-22	14 Mar 1998–09 Apr 1998	F/A-18C(N)	CVN-70	CVW-11	NH-3xx	EastPac
VFA-22	22 Jun 1998–11 Aug 1998	F/A-18C(N)	CVN-70	CVW-11	NH-3xx	JTFEX/FLEETEX, RIMPAC 98
VFA-22	06 Nov 1998–06 May 1999	F/A-18C(N)	CVN-70	CVW-11	NH-3xx	WestPac, Persian Gulf
VFA-22	23 Jul 2001–23 Jan 2002	F/A-18C(N)	CVN-70	CVW-11	NH-3xx	WestPac, Persian Gulf
VFA-22	13 Jan 2003–19 Sep 2003	F/A-18C(N)	CVN-70	CVW-9	NG-1xx	WestPac

VFA-22 has served predominantly within the structure of CVW-11 while operating the legacy version of the Hornet. Here 164060 is seen in November 2003 shortly after its final cruise before the jets were traded in for the more advanced Super Hornet. The aircraft is currently assigned to the US Marine Corps with VMFAT-101.

VFA-25 Fist of the Fleet

In May 1983, VA-25 began transition to the new Boeing F/A-18A Hornet and was re-designated as Strike Fighter Squadron 25 (VFA-25) on 1 July 1983. On 19 February 1985, VFA-25 departed on the first deployment of the F/A-18 Hornet aboard America's Flagship, the USS *Constellation* (CV-64). During this WestPac deployment, VFA-25 won both the CVW-14 Bombing Derby and the CVW-14 Tailhook Award. The squadron came back to NAS Lemoore in August 1985, then undertook four further deployments aboard the *Connie* before cross-decking to CV-62 for a RIMPAC exercise in April 1990.

After a further WestPac cruise with the A model, the squadron transitioned to the newer C for its final short cruise aboard CV-62, which ended up in Hawaii. From here VFA-25 was reunited with the A model again for its return to the United States aboard CV-41, USS *Midway*, on her final cruise.

Before the squadron's next deployment in February 1994 it once again upgraded to the C model but this time the night-attack version, which it took to sea aboard CVN-70 for the first of three cruises aboard the ship. The squadron were, according to *Go Navy*, to undertake seventeen cruises with the night-attack version of the F/A-18C participating in Operation Southern Watch, Operation Desert Strike and exercise Rugged Nautilus.

The cruise began on 19 March 2003 and saw the squadron once again begin combat sorties in earnest in support of Operation Southern Watch, which quickly became Operation Iraqi Freedom, and it was involved in important and urgent attacks that played a role of vital importance in the demise of the rogue Iraqi administration. VFA-25 delivered more than 300,000lb of ordnance during this operation.

Another six-month WestPac deployment began in 2006, this one being the squadron's first aboard the USS *Ronald Reagan* (CVN-76). VFA-25 performed nearly four months of combat operations in the Persian Gulf from February to May 2006, during which time it successfully carried out upwards of 600 combat sorties and flew 2,500 hours in support of Operation Iraqi Freedom.

The rest of 2006 was spent in a 'sustainment status', as the squadron prepared for a three-month deployment due to begin in January 2007, followed by the Strike Fighter Advanced Readiness Program (SFARP) in September 2007 as the squadron began preparations for a further deployment cycle.

VFA-25 had seen almost a full year at sea in a period of less than four years overall. During this time it participated in three full and one surge deployments and logged more than 5,500 combat hours on nearly 1,000 sorties in both Iraq and Afghanistan.

VFA-25 re-equipped with the F/A-18A in 1983 before upgrading to the more advanced C model in 1990. However, it swapped these for the A model for the final cruise of the USS *Midway*, although was reunited with the C once again upon return to the United States. Aircraft 162854 is seen in June 1988 while part of the USS *Constellation* air wing. It now serves with the reserves of VFA-204. (Photo Don Jay)

In autumn 2010 the squadron returned aboard CVN-70 for what was to be the final two deployments with its aging C models, then in 2012 VFA-25 completed its transition to the F/A-18E Super Hornet and joined Carrier Air Wing Nine (CVW-9) on board USS *John C. Stennis* (CVN-74).

VFA-25	20 Feb 1985–24 Aug 1985	F/A-18A	CV-64	CVW-14	NK-4xx	WestPac, Indian Ocean
VFA-25	04 Sep 1986–20 Oct 1986	F/A-18A	CV-64	CVW-14	NK-4xx	NorPac
VFA-25	11 Apr 1987–13 Oct 1987	F/A-18A	CV-64	CVW-14	NK-4xx	WestPac, Indian Ocean, Arabian Gulf
VFA-25	01 Dec 1988–01 Jun 1989	F/A-18A	CV-64	CVW-14	NK-4xx	WestPac, Indian Ocean
VFA-25	16 Sep 1989–19 Oct 1989	F/A-18A	CV-64	CVW-14	NK-4xx	NorPac
VFA-25	27 Apr 1990–20 May 1990	F/A-18A	CV-62	CVW-14	NK-4xx	RIMPAC
VFA-25	23 Jun 1990–20 Dec 1990	F/A-18A	CV-62	CVW-14	NK-4xx	WestPac, Persian Gulf
VFA-25	05 Aug 1991–22 Aug 1991	F/A-18C	CV-62	CVW-14	NK-4xx	west coast to Hawaii
VFA-25	28 Aug 1991–14 Sep 1991	F/A-18A	CV-41	CVW-14	NK-4xx	Hawaii to west coast
VFA-25	17 Feb 1994–17 Aug 1994	F/A-18C(N)	CVN-70	CVW-14	NK-4xx	WestPac, Persian Gulf
VFA-25	21 Aug 1995–13 Sep 1995	F/A-18C(N)	CVN-70	CVW-14	NK-4xx	Hawaii Operating Area
VFA-25	14 May 1996–14 Nov 1996	F/A-18C(N)	CVN-70	CVW-14	NK-4xx	WestPac, Persian Gulf
VFA-25	04 Apr 1997–28 Apr 1997	F/A-18C(N)	CVN-72	CVW-14	NK-4xx	EastPac
VFA-25	22 Nov 1997–20 Dec 1997	F/A-18C(N)	CVN-72	CVW-14	NK-4xx	EastPac
VFA-25	11 Jun 1998–11 Dec 1998	F/A-18C(N)	CVN-72	CVW-14	NK-4xx	WestPac, Persian Gulf
VFA-25	12 May 2000–01 Jul 2000	F/A-18C(N)	CVN-72	CVW-14	NK-4xx	RIMPAC 2000
VFA-25	17 Aug 2000–12 Feb 2001	F/A-18C(N)	CVN-72	CVW-14	NK-4xx	WestPac, Persian Gulf
VFA-25	14 Apr 2002–14 May 2002	F/A-18C(N)	CVN-72	CVW-14	NK-4xx	NorPac, JTFEX

VFA-25	20 Jul 2002–06 May 2003	F/A-18C(N)	CVN-72	CVW-14	NK-4xx	WestPac, Persian Gulf
VFA-25	24 May 2004–01 Nov 2004	F/A-18C(N)	CVN-74	CVW-14	NK-4xx	RIMPAC 04, WestPac
VFA-25	04 Jan 2006–06 Jul 2006	F/A-18C(N)	CVN-76	CVW-14	NK-4xx	WestPac, Persian Gulf
VFA-25	27 Jan 2007–20 Apr 2007	F/A-18C(N)	CVN-76	CVW-14	NK-4xx	WestPac
VFA-25	19 May 2008–25 Nov 2008	F/A-18C(N)	CVN-76	CVW-14	NK-4xx	WestPac, north Arabian Sea
VFA-25	28 May 2009–21 Oct 2009	F/A-18C(N)	CVN-76	CVW-14	NK-4xx	WestPac, north Arabian Sea
VFA-25	30 Nov 2010–15 Jun 2011	F/A-18C(N)	CVN-70	CVW-17	AA-4xx	COMPTUEX, WestPac, north Arabian Sea
VFA-25	30 Nov 2011–23 May 2012	F/A-18C(N)	CVN-70	CVW-17	AA-4xx	WestPac, CENTCOM AOR

VFA-27 Royal Maces

On 21 January 1991, VA-27, the Royal Maces, transitioned to the F/A-18A Hornet and was re-designated Strike Fighter Squadron 27, the Chargers. This was after a hugely successful twenty-three years on the LTV A-7A/E Corsair II, encompassing fourteen deployments, combat hours numbering well into the thousands and many unit awards. VA-27, based at NAS Lemoore, was the last A-7E Corsair II squadron to transition to the Hornet.

In November 1992, VFA-27 deployed with Lot 8 Hornets aboard USS *Kitty Hawk* (CV-63) as part of the CVW-15 'Wolfpack' for a Western Pacific deployment participating off the coast of Somalia in support of Operation Restore Hope. The squadron later flew to Dhahran, Saudi Arabia, to enhance the United States Central Command's Air Forces Coalition supporting Operation Southern Watch. On the night of 13 January 1993 the squadron undertook a strike against Iraq, delivering more than 18,000lb of ordnance.

In June 1994, the squadron again deployed as part of Wing 15 aboard the 'Hawk' for the Western Pacific. By December 1994, the unit had completed its sixteenth Western Pacific (WestPac) cruise and then began transition to Lot 12 F/A-18Cs. It also started getting ready to change its home port to Atsugi, Japan, which took place in 1996. On 4 June the Chargers name was abandoned altogether and the squadron resumed its original designation as the Royal Maces. Simultaneously, the Royal Maces TransPac'd all twelve squadron aircraft from NAS Lemoore to NAF Atsugi to join the CVW-5/USS *Independence* (CV-62) 'I-5' Team.

VFA-27 operated the A model between 1991 and 1994, when it transitioned to the night-attack version of the F/A-18C(N). Here can be seen F/A-18A 162422 sporting the squadron markings in 1991 shortly after delivery. The jet has subsequently been stricken from the inventory and was last noted in store at NAS North Island in 2002. (Photo Jerry Geer)

The squadron was to undertake fifteen deployments with the night-attack version of the F/A-18C version of the Hornet and in this period operated in support of Operation Southern Watch as well as numerous exercises including RIMPAC 98. The unit also positively deployed the Joint Stand-off Weapon (JSOW), in the process showcasing its unique capabilities.

In the second week of April 2000 the squadron boarded the USS *Kitty Hawk* (CV-63) in Guam, and then participated in the Tandem Thrust 2000 exercise, Cobra Gold, followed by Foal Eagle, Cope Thunder and Annualex 2000, before finally returning home in November.

Following the terrorist attacks on 11 September 2001, the Royal Maces took part in in Operation Enduring Freedom, flying missions against the Al-Qaeda infrastructure and Taliban forces in Afghanistan as well as defending important assets in Diego Garcia.

When *Kitty Hawk* deployed to the Persian Gulf in 2003 in support of Operation Iraqi Freedom, VFA-27 soon played a significant role. As part of Carrier Air Wing Five (CVW-5), VFA-27 attacked command, control and communications sites, surface-to-surface missile batteries and also an air traffic control radar near Basra. The squadron flew more than 300 close air support and strike sorties and used more than 200,000lb of ordnance against Iraqi forces in support of Operation Iraqi Freedom, achieving a hitherto unprecedented 100 per cent sortie completion rate.

The Royal Maces of Strike Fighter Squadron 27 left USS *Kitty Hawk* (CV-63) on 24 May 2004 and began a transition from the F/A-18C(N) Hornet to the F/A-18E Super Hornet.

VFA-27	18 Oct 1991–11 Dec 1991	F/A-18A	CV-63	CVW-15	NL-4xx	Norfolk to San Diego
VFA-27	03 Nov 1992–03 May 1993	F/A-18A	CV-63	CVW-15	NL-4xx	WestPac, Persian Gulf
VFA-27	24 Jun 1994–22 Dec 1994	F/A-18A	CV-63	CVW-15	NL-4xx	WestPac, Indian Ocean
VFA-27	15 Feb 1997–10 Jun 1997	F/A-18C(N)	CV-62	CVW-5	NF-2xx	WestPac, Indian Ocean
VFA-27	23 Jan 1998–05 Jun 1998	F/A-18C(N)	CV-62	CVW-5	NF-2xx	WestPac, Persian Gulf
VFA-27	Jul 1998–17 Jul 1998	F/A-18C(N)	CV-62	CVW-5	NF-2xx	Yokosuka to Hawaii
VFA-27	24 Jul 1998 – 11 Aug 1998	F/A-18C(N)	CV-63	CVW-5	NF-2xx	Hawaii to Yokosuka
VFA-27	30 Sep 1998–13 Nov 1998	F/A-18C(N)	CV-63	CVW-5	NF-2xx	WestPac, Sea of Japan
VFA-27	02 Mar 1999–25 Aug 1999	F/A-18C(N)	CV-63	CVW-5	NF-2xx	WestPac, Persian Gulf
VFA-27	22 Oct 1999–10 Nov 1999	F/A-18C(N)	CV-63	CVW-5	NF-2xx	WestPac, Sea of Japan
VFA-27	11 Apr 2000–05 Jun 2000	F/A-18C(N)	CV-63	CVW-5	NF-2xx	WestPac
VFA-27	26 Sep 2000–20 Nov 2000	F/A-18C(N)	CV-63	CVW-5	NF-2xx	WestPac, Sea of Japan
VFA-27	02 Mar 2001–11 Jun 2001	F/A-18C(N)	CV-63	CVW-5	NF-2xx	WestPac
VFA-27	15 Apr 2002–05 Jun 2002	F/A-18C(N)	CV-63	CVW-5	NF-2xx	WestPac
VFA-27	25 Oct 2002–13 Dec 2002	F/A-18C(N)	CV-63	CVW-5	NF-2xx	WestPac
VFA-27	23 Jan 2003–06 May 2003	F/A-18C(N)	CV-63	CVW-5	NF-2xx	WestPac, Persian Gulf
VFA-27	01 Nov 2003–12 Dec 2003	F/A-18C(N)	CV-63	CVW-5	NF-2xx	WestPac
VFA-27	18 Feb 2004–24 May 2004	F/A-18C(N)	CV-63	CVW-5	NF-2xx	WestPac

VFA-34 Blue Blasters

On 30 September 1996, Attack Squadron 34 (VA-34) was re-designated Strike Fighter Squadron 34 (VFA-34), giving up its Grumman A-6E Intruder aircraft and returning once again to NAS Cecil Field, Florida, where it began the transition to the F/A-18C(N) Hornet. VFA-34 deployed aboard USS *Dwight D. Eisenhower* (CVN-69) in June 1998, this deployment occurring as part of Carrier Air Wing Seventeen (CVW-17) in support of Operation Deliberate Force and Operation Southern Watch.

The squadron undertook four deployments as part of CVW-17 and a further nine as part of Air Wing Two operating from six different aircraft carriers. Today the squadron is one of only four 'fleet' squadrons to still operate the legacy Hornet.

VFA-34 is one of the few US Navy fleet squadrons still operating the C model. Here a pair of jets are seen getting airborne from NAS Oceana. (Photo Scott Rathbone)

VFA-34	10 Jun 1998–10 Dec 1998	F/A-18C(N)	CVN-69	CVW-17	AA-2xx	Med
VFA-34	21 Jun 2000–19 Dec 2000	F/A-18C(N)	CVN-73	CVW-17	AA-2xx	Med, Persian Gulf
VFA-34	20 Jun 2002–20 Dec 2002	F/A-18C(N)	CVN-73	CVW-17	AA-2xx	Med, north Arabian Sea
VFA-34	07 Jun 2004–13 Dec 2004	F/A-18C(N)	CV-67	CVW-17	AA-2xx	Med, Persian Gulf
VFA-34	01 Jun 2005–23 Jun 2005	F/A-18C(N)	CVN-72	CVW-2	NE-4xx	NorPac
VFA-34	19 Oct 2005–16 Nov 2005	F/A-18C(N)	CVN-72	CVW-2	NE-4xx	EastPac
VFA-34	27 Feb 2006–08 Aug 2006	F/A-18C(N)	CVN-72	CVW-2	NE-4xx	WestPac
VFA-34	13 Mar 2008–12 Oct 2008	F/A-18C(N)	CVN-72	CVW-2	NE-4xx	WestPac, Persian Gulf
VFA-34	07 Sep 2010–24 Mar 2011	F/A-18C(N)	CVN-72	CVW-2	NE-4xx	WestPac, north Arabian Sea
VFA-34	07 Dec 2011–07 Aug 2012	F/A-18C(N)	CVN-72	CVW-2	NE-4xx	World Cruise
VFA-34	13 Jun 2014–10 Aug 2014	F/A-18C(N)	CVN-76	CVW-2	NE-4xx	RIMPAC
VFA-34	08 Sep 2015–17 Dec 2015	F/A-18C(N)	CVN-73	CVW-2	NE-4xx	San Diego to Norfolk
VFA-34	02 Feb 2017– 23 Jun 2017	F/A-18C(N)	CVN-70	CVW-2	NE-4xx	WestPac
VFA-34	01 Jan 2018 – 11 Apr 2018	F/A-18C(N)	CVN-70	CVW-2	NE-4xx	WestPac

VFA-37 Ragin' Bulls

On 31 October 1990, the Bulls flew their final official sortie in the A-7E Corsair II, thus ending a twenty-five-year association in which more than 115,000 flight hours and 25,000 arrested landings were accumulated.

On 15 November 1990 Attack Squadron Three Seven was officially re-designated Strike Fighter Squadron Three Seven (VFA-37), receiving its first F/A-18C Hornet on 13 December 1990 and on 1 September 1991 was attached to Carrier Air Wing Three (CVW-3), embarking on the USS *John F. Kennedy* (CV-67).

The squadron has undertaken twelve deployments with the night-attack version of the Boeing F/A-18C Hornet, predominantly within the structure of CVW-3. VFA-37's latest cruise saw it join CVW-8, taking the place of VFA-15, which had recently been disestablished. The squadron was one of only four fleet squadrons still operating the legacy Hornet at the end of 2017.

The VFA-37 Ragin' Bulls commander's aircraft, 165203, is seen turning finals at NAS Oceana. The aircraft, sporting the authorised one special scheme per squadron, still serves in the same capacity, having flown continuously within the structure of Air Wing 3 since delivery.

The deployments saw the squadron support a number of ongoing operations, including Operation Provide Comfort in northern Iraq and Operation Provide Promise near the former Yugoslavia, as well as later over Bosnia in support of Operation Deny Flight.

The Bulls embarked on board USS *Theodore Roosevelt* (CVN-71) in November 1996 as part of CVW-3, flying in support of Operation Southern Watch in Iraq and Operation Deliberate Guard over Bosnia-Herzegovina.

The executive order issued for Operation Desert Fox in Iraq saw the squadron fly forty-four combat sorties and deliver 46 tons of ordnance in four nights of sustained combat operations. The hit assessment for VFA-37 was apparently the best in the air wing and the operation was completed with no friendly losses.

The Ragin' Bulls relocated from NAS Cecil Field, Florida, to NAS Oceana, Virginia in July 1999 as part of the Base Realignment and Closure (BRAC) programme. The next major deployment for the Bulls was on the maiden voyage of the USS *Harry S. Truman* (CVN-75) in 2000.

In December 2002, for its second cruise aboard the *Truman*, the squadron joined other elements of the fleet spearheading combat operations in Operation Iraqi Freedom, participating in the 'shock and awe' campaign in southern Iraq before moving on to assist US forces in northern Iraq.

After thirty days, the Bulls had flown upwards of 1,200 hours, with 252 combat sorties over northern Iraq. The squadron pilots delivered more than 144 tons of ordnance and expended 9,400 rounds of 20mm HEI (High Explosive) ammunition.

The Ragin' Bulls enjoyed their fortieth anniversary in 2007. VFA-37 began preparations with CVW-3 for another deployment, following on from sustainment operations with Carrier Air Wing Eight (CVW-8) the preceding year. The Truman strike group left Virginia on 5 November 2007 to provide support for Operations Enduring Freedom and Iraqi Freedom. The squadron returned in June 2008.

After returning from the seven-month deployment, the Bulls underwent a period of sustainment and have subsequently undertaken two further Mediterranean and north Arabian Sea deployments as part of CVN-75 and CVW-3. They cross-decked to the USS *George H.W. Bush* (CVN-77) in 2017, taking the place of the disbanded VFA-15 within CVW-8.

VFA-37	7 Oct 1992–07 Apr 1993	F/A-18C(N)	CV-67	CVW-3	AC-3xx	Med
VFA-37	20 Oct 1994–14 Apr 1995	F/A-18C(N)	CVN-69	CVW-3	AC-3xx	Med
VFA-37	25 Nov 1996–22 May 1997	F/A-18C(N)	CVN-71	CVW-3	AC-3xx	Med
VFA-37	06 Nov 1998–06 May 1999	F/A-18C(N)	CVN-65	CVW-3	AC-3xx	Med, Adriatic Sea, Persian Gulf
VFA-37	28 Nov 2000–23 May 2001	F/A-18C(N)	CVN-75	CVW-3	AC-3xx	Med, Persian Gulf
VFA-37	05 Dec 2002–23 May 2003	F/A-18C(N)	CVN-75	CVW-3	AC-3xx	Med
VFA-37	02 Jun 2004–25 Jul 2004	F/A-18C(N)	CVN-75	CVW-3	AC-3xx	Lant, Med
VFA-37	13 Oct 2004–18 Apr 2005	F/A-18C(N)	CVN-75	CVW-3	AC-3xx	Med, Persian Gulf
VFA-37	05 Nov 2007–04 Jun 2008	F/A-18C(N)	CVN-75	CVW-3	AC-3xx	Med, Persian Gulf
VFA-37	21 May 2010–20 Dec 2010	F/A-18C(N)	CVN-75	CVW-3	AC-3xx	Med, north Arabian Sea
VFA-37	22 Jul 2013–18 Apr 2014	F/A-18C(N)	CVN-75	CVW-3	AC-3xx	Med, north Arabian Sea
VFA-37	21 Jan 2017–22 Aug 2017	F/A-18C(N)	CVN-77	CVW-8	AJ-4xx	Med, North Arabian Sea

VFA-81 Sunliners

Transition from the venerable LTV A-7E Corsair II to the new Boeing F/A-18C Hornet saw VA-81 re-designated VFA-81 on 4 February 1988.

Two years later, the squadron made the first of three deployments aboard USS *Saratoga* (CV-60) between 1990 and 1994. On its first cruise operating the Hornet, VFA-81 took part in the Gulf War. On 17 January 1991, Lt Cdr Scott Speicher was shot down and killed by a Mig-25PDS 'Foxbat' piloted by Lt Zuhair Dawood, 84th Fighter Squadron of the Iraqi Air Force, while flying a mission west of Baghdad. This occurred on the very first night of the war. Also on the first day of the war, the squadron scored the Navy's only two aerial victories over enemy fighters during the campaign by downing two Iraqi MiG-21s.

Seen at Homestead AFB in October 1988, F/A-18C 163480 had just been delivered to VFA-81. The unit assigned to CVW-17 converted directly onto the C model, a mount it was to retain for twenty years. This particular aircraft is still in the inventory and was last noted in store at NAS North Island.

VFA-81 returned to its home port of NAS Cecil Field, Florida, on 27 March 1991. Later the squadron also participated in the last Mediterranean deployment of USS *Saratoga*, which was decommissioned in August 1994.

Two cruises followed the first in 1996 with CVW-17 aboard the 'Big E', USS *Enterprise* (CVN-65), followed by a deployment aboard the USS *Dwight D. Eisenhower* in 1998 (CVN-69). Following this cruise, the squadron shifted its home port to NAS Oceana, Virginia, due to the BRAC-mandated closure of NAS Cecil Field in 1999.

In 2000 and 2002 VFA-81 made two deployments aboard USS *George Washington* (CVN-73) to the Mediterranean Sea and the Persian Gulf prior to 2004 when the squadron made a single deployment aboard the USS *John F. Kennedy* CV-67) as she was coming to the end of her useful life.

By 2007 VFA-81 was reassigned to CVW-11 in the Pacific Fleet and deployed aboard USS *Nimitz* (CVN-68) to the Western Pacific and the Persian Gulf. Its second and last deployment ending on 3 June 2008, after which the squadron began transition training from the F/A-18C Hornet to the F/A-18E Super Hornet and in the process returning to CVW-17.

VFA-81	07 Aug 1990–28 Mar 1991	F/A-18C	CV-60	CVW-17	AA-4xx	Med, Red Sea
VFA-81	06 May 1992–06 Nov 1992	F/A-18C	CV-60	CVW-17	AA-4xx	Med
VFA-81	12 Jan 1994–24 Jun 1994	F/A-18C	CV-60	CVW-17	AA-4xx	Med
VFA-81	21 Feb 1996–01 Apr 1996	F/A-18C	CVN-65	CVW-17	AA-4xx	NorLant
VFA-81	28 Jun 1996–20 Dec 1996	F/A-18C	CVN-65	CVW-17	AA-4xx	Med
VFA-81	10 Jun 1998–10 Dec 1998	F/A-18C	CVN-69	CVW-17	AA-4xx	Med
VFA-81	21 Jun 2000–19 Dec 2000	F/A-18C	CVN-73	CVW-17	AA-4xx	Med, Persian Gulf
VFA-81	20 Jun 2002–20 Dec 2002	F/A-18C	CVN-73	CVW-17	AA-4xx	Med, north Arabian Sea
VFA-81	07 Jun 2004–13 Dec 2004	F/A-18C	CV-67	CVW-17	AA-4xx	Med, Persian Gulf
VFA-81	04 Apr 2006–24 May 2006	F/A-18C	CVN-73	CVW-17	AA-4xx	Caribbean Sea
VFA-81	02 Apr 2007–30 Sep 2007	F/A-18C	CVN-68	CVW-11	NH-4xx	WestPac, Persian Gulf
VFA-81	24 Jan 2008–03 Jun 2008	F/A-18C	CVN-68	CVW-11	NH-4xx	WestPac

VFA-82 Marauders

VA-82 was re-designated Strike Fighter Squadron 82 (VFA-82) as the first Hornet aircraft was delivered in November 1987. At that time VFA-82 became the first squadron to operate the F/A-18C model, having completed its last cruise with the LTV A-7E Corsair II in June 1987.

In 1990, VFA-82 was part of the air wing aboard the USS Constellation as she made her way around the horn of South America on the way to Philadelphia for upgrading. December 1990 found VFA-82 aboard the USS *America* (CV-66) as part of CVW-1, participating in Operation Desert Shield and Operation Desert Storm. The Marauders flew 597 combat sorties, 1,308.9 combat hours and delivered more than 1.2 million pounds of ordnance over Iraq during this conflict.

In December 1993, the squadron returned to the USS *America* for a six-month deployment, during which it participated in Operation Southern Watch in Iraq, Operation Deny Flight and Operation Sharp Guard in Bosnia and Operation Restore Hope in Somalia.

The unit also joined the USS *America* for her last cruise undertaken in the Mediterranean and Persian Gulf from August 1995 to February 1996 following which, along with the rest of the wing, it deployed on board USS *George Washington* (CVN-73) to the Mediterranean Sea and Persian Gulf in support of Operation Southern Watch.

During 1998 the VFA-82 operated from the decks of three different US aircraft carriers, CVN-73, 75 and CV-67. The squadron also went to seven different countries during this period, having by this time upgraded to the night-attack version of the F/A-18C(N).

In August 1999, the squadron left NAS Cecil Field and relocated temporarily to NAS Jacksonville for two months prior to deployment aboard the USS *John F. Kennedy* (CV-67) in support of Operation Southern Watch. After this six-month deployment was complete, the Marauders moved to MCAS Beaufort, South Carolina.

Another squadron to miss out on the Alpha model Hornet was VFA-82. Converting directly to the C model, the squadron deployed aboard the USS *America* and participated in both Operations Desert Shield and Desert Storm. This jet, 165203, is seen here taxiing at NAS New Orleans in October 2000 and remains in the active inventory.

During four months of combat operations, VFA-82 was to deliver more than 440,000lb of ordnance and totalled more than 3,000 hours in support of Operation Enduring Freedom. The unit returned to Beaufort on 28 March 2002 having logged a record 159 days continuously deployed. Remarkably, most of this time was in combat operations.

Cross-decking once again, CVW-1 set sail in CVN-65 in August 2003, again to the Mediterranean and Persian Gulf, and returned in June 2004 before a quick turn for a short Atlantic cruise the following month. The squadron's last deployment took place in October 2004 aboard CVN-72 as part of CVW-2 on its only WestPac cruise with the Hornet, which lasted until March 2005. Upon return VFA-82 was disestablished, this action becoming effective on 30 September 2005, and following on from a directive issued on 5 July.

VFA-82	21 Mar 1988–08 May 1988	F/A-18C	CV-66	CVW-1	AB-3xx	SoLant
VFA-82	08 Feb 1989–03 Apr 1989	F/A-18C	CV-66	CVW-1	AB-3xx	NorLant
VFA-82	11 May 1989–10 Nov 1989	F/A-18C	CV-66	CVW-1	AB-3xx	Med, Indian Ocean
VFA-82	28 Dec 1990–18 Apr 1991	F/A-18C	CV-66	CVW-1	AB-3xx	Arabian Gulf
VFA-82	21 Aug 1991–11 Oct 1991	F/A-18C	CV-66	CVW-1	AB-3xx	NorLant
VFA-82	02 Dec 1991–06 Jun 1992	F/A-18C	CV-66	CVW-1	AB-3xx	Med
VFA-82	11 Aug 1993–05 Feb 1994	F/A-18C	CV-66	CVW-1	AB-3xx	Med
VFA-82	28 Aug 1995–24 Feb 1996	F/A-18C	CV-66	CVW-1	AB-3xx	Med, Persian Gulf
VFA-82	03 Oct 1997–03 Apr 1998	F/A-18C(N)	CVN-73	CVW-1	AB-3xx	Med, Persian Gulf
VFA-82	02 Nov 1998–17 Dec 1998	F/A-18C(N)	CVN-75	CVW-1	AB-3xx	Puerto Rican Operating Area
VFA-82	22 Sep 1999–19 Mar 2000	F/A-18C(N)	CV-67	CVW-1	AB-3xx	Med, Persian Gulf
VFA-82	19 Sep 2001–27 Mar 2002	F/A-18C(N)	CVN-71	CVW-1	AB-3xx	Med, Northern Arabian Sea
VFA-82	28 Aug 2003–29 Feb 2004	F/A-18C(N)	CVN-65	CVW-1	AB-3xx	SoLant, Med, Persian Gulf
VFA-82	03 Jun 2004–23 Jul 2004	F/A-18C(N)	CVN-65	CVW-1	AB-3xx	EastLant, NorLant
VFA-82	15 Oct 2004–04 Mar 2005	F/A-18C(N)	CVN-72	CVW-2	NE-4xx	WestPac

VFA-83 Rampagers

The year 1987 marked VA-83's final deployment with the LTV A-7E Corsair II, and in November it began the transition to the Boeing F/A-18C Hornet, with the unit re-designated as Strike Fighter Squadron (VFA-83) on 1 March 1988.

In mid 1990 the squadron made its first Hornet deployment aboard the USS *Saratoga* (CV-60) in support of Operation Desert Shield and Operation Desert Storm. During the forty-three days of operations it flew some 237 combat missions over Iraq and Kuwait.

In 1992 the squadron was deployed to the Adriatic Sea supporting United Nations operations in the former Yugoslavia, then in 1994 a second deployment to the area marked the first ever detachment to Jordan. After three cruises aboard the old *Saratoga*, the squadron once again deployed to the Mediterranean Sea, the Adriatic Sea and the Persian Gulf but this time on board the USS *Enterprise* (CVN-65) in support of Operation Southern Watch.

Following this deployment, the squadron then transitioned to the more capable F/A-18C(N) version of the Hornet before in April 1998 relocating its home base to NAS Oceana, Virginia. During the same year the squadron was deployed aboard the USS *Dwight D. Eisenhower* (CVN-69) to the Mediterranean Sea.

Another early block 23 C model Hornet bearing the squadron commander's markings in June 1992. 163444 went on to serve with the west coast Replacement Air Squadron (RAS) at NAS Lemoore. (Photo via Andy Thomson)

The squadron, along with the rest of the wing, CVW-17, was then reassigned to the carrier USS *George Washington* (CVN-73) and made two deployments to the Mediterranean Sea and the Arabian Sea in 2000 and 2002, supporting Operation Enduring Freedom and Operation Iraqi Freedom. These operations continued in 2004 with the squadron embarked on the last deployment of the *USS John F. Kennedy* (CV-67).

VFA-83 was then temporarily reassigned to Carrier Wing Seven (CVW-7) and deployed again on the USS *Dwight D. Eisenhower* to both the Mediterranean and the Indian Ocean during late 2006 and early in 2007. With the USS *Dwight D Eisenhower* entering a refit in 2008, the strike fighter squadrons of CVW-7 were re-assigned to CVW-17 and conducted a period of preparation on the USS *George Washington*, during which time CVW-7's squadrons kept their tail code 'AG'.

The squadron, and wing, undertook four cruises to the Mediterranean and North Africa with the 'GW', the last ending in July 2013. The squadron's next cruise was in November 2015 when it joined CVN-75 as part of the Fifth Fleet. Until recently the squadron, and CVW-7, were assigned to the USS *Dwight D. Eisenhower* (CVN-69), although they may well transfer to the US Navy's newest carrier, CVN-78, when the unit transitions to a newer platform.

VFA-83	07 Aug 1990–28 Mar 1991	F/A-18C	CV-60	CVW-17	AA-3xx	Med, Red Sea
VFA-83	06 May 1992–06 Nov 1992	F/A-18C	CV-60	CVW-17	AA-3xx	Med
VFA-83	12 Jan 1994–24 Jun 1994	F/A-18C	CV-60	CVW-17	AA-3xx	Med
VFA-83	21 Feb 1996 –01 Apr 1996	F/A-18C	CVN-65	CVW-17	AA-3xx	NorLant
VFA-83	28 Jun 1996–20 Dec 1996	F/A-18C	CVN-65	CVW-17	AA-3xx	Med
VFA-83	10 Jun 1998–10 Dec 1998	F/A-18C(N)	CVN-69	CVW-17	AA-3xx	Med
VFA-83	21Jun 2000–19 Dec 2000	F/A-18C(N)	CVN-73	CVW-17	AA-3xx	Med, Persian Gulf
VFA-83	20 Jun 2002–20 Dec 2002	F/A-18C(N)	CVN-73	CVW-17	AA-3xx	Med, north Arabian Sea
VFA-83	07 Jun 2004–13 Dec 2004	F/A-18C(N)	CV-67	CVW-17	AA-3xx	Med, Persian Gulf
VFA-83	03 Oct 2006–23 May 2007	F/A-18C(N)	CVN-69	CVW-7	AG-3xx	Med, Persian Gulf
VFA-83	07 Apr 2008–27 May 2008	F/A-18C(N)	CVN-73	CVW-17	AG-3xx	Norfolk to San Diego
VFA-83	21 Feb 2009–30 Jul 2009	F/A-18C(N)	CVN-69	CVW-7	AG-3xx	Med, north Arabian Sea
VFA-83	02 Jan 2010–28 Jul 2010	F/A-18C(N)	CVN-69	CVW-7	AG-3xx	Med, north Arabian Sea
VFA-83	20 Jun 2012–19 Dec 2012	F/A-18C(N)	CVN-69	CVW-7	AG-3xx	Med, north Arabian Sea, Persian Gulf
VFA-83	21 Feb 2013–03 Jul 2013	F/A-18C(N)	CVN-69	CVW-7	AG-3xx	Med, north Arabian Sea
VFA-83	16 Nov 2015–13 Jul 2016	F/A-18C(N)	CVN-75	CVW-7	AG-3xx	Med, 5th Fleet AOR

VFA-86 Sidewinders

VA-86 was the first east coast squadron to receive the C model (Lot 10) F/A-18 Hornet and was officially re-designated as Strike Fighter Squadron 86 (VFA-86) on 15 July 1987. From 1988 to 2007 the squadron was assigned to Carrier Air Wing One (CVW-1). The first cruise in 1988, an eight-month deployment, was aboard the USS *America*, during which time the wing provided air cover for the evacuation of the US Embassy in Beirut.

In 1991 the squadron flew combat missions in support of Operation Desert Storm from USS *America* operating in both the Red Sea and the Persian Gulf. Upon returning to the USA in 1992, the squadron accomplished the first F/A-18 Standoff Land Attack Missile (SLAM) shoot by an east-coast squadron. This was to a target on San Nicolas Island in California.

F/A-18C 163443 'AB-400' of VFA-86 seen in September 1992. This aircraft later went on to serve with the US Navy's display team, the Blue Angels. It was retired to AMARG from VA-125 on 22 October 2009. Identified as a suitable candidate for the return to service programme, it departed for the Boeing facility at Cecil Field, Florida, on 29 August 2016. (Photo via Geoff Rhodes)

Deployment aboard *America* was undertaken again in 1993 in support of Operation Deny Flight but in October the carrier was ordered to the Indian Ocean off the coast of Somalia as part of the relief efforts supporting the United Nations. Then, after conducting NATO Operation Deliberate Force missions over Bosnia-Herzegovina from the Adriatic Sea, the squadron returned to its base at NAS Cecil Field, Florida, in February 1994. A further deployment aboard the USS *America* followed in 1995 before the squadron retuned to the United States in February 1996 and said farewell to USS *America* for the last time.

On 3 October 1997, the squadron, as part of CVW-1, deployed to the Mediterranean Sea/Persian Gulf aboard USS *George Washington* (CVN-73). The Sidewinders remained on station in the Gulf through to the middle of March 1998. This was the crisis period of Iraq's non-compliance with UN sanctions. The squadron flew Operation Southern Watch contingency missions.

In November to December 1998, VFA-86 made a short cruise to the Puerto Rican Operating Area. This trip was aboard CVN-75, USS *Harry S. Truman*, and was followed by a further deployment to the Mediterranean Sea/Arabian Sea, this time on the USS *John F. Kennedy*, participating in Freedom of Navigation Operations off the Libyan coast.

Following the decommissioning of NAS Cecil Field in 1999 due to the BRAC programme, VFA-86 was one of two US Navy strike fighter squadrons that relocated to MCAS Beaufort, South Carolina, in lieu of NAS Oceana, Virginia. The other Navy squadron to transfer to MCAS Beaufort, VFA-82, later deactivated in 2005.

CVW-1 cross-decked once again in 2001 when it was deployed aboard CVN-71, USS *Theodore Roosevelt*, for one cruise lasting until March 2002 in support of Operation Enduring Freedom over Afghanistan against Taliban and Al-Qaeda military targets. When it left the area on 2 March 2002, the squadron had flown more than 3,500 hours and used more than 213 tons of ordnance. It had been at sea for 160 continuous days.

Remaining with the air wing, VFA-86 moved over to USS *Enterprise* (CVN-65) in support of Operations Iraqi Freedom and Enduring Freedom for two cruises as part of the Sixth Fleet before transition to the night interdiction role with the F/A-18C(N). The first of these saw the squadron operate throughout the Mediterranean and Middle East before ending up in the Pacific. There followed a further cruise in 2007 aboard the 'Big E' before the unit was re-assigned to CVW-11 on the west coast, in the process joining USS *Nimitz*, CVN-68.

This cruise ended in March 2010, after which the squadron relocated from MCAS Beaufort to NAS Lemoore and stood down for conversion to the F/A-18E Super Hornet.

VFA-86	21 Mar 1988–08 May 1988	F/A-18C	CV-66	CVW-1	AB-4xx	SoLant
VFA-86	08 Feb 1989 – 03 Apr 1989	F/A-18C	CV-66	CVW-1	AB-4xx	NorLant
VFA-86	11 May 1989–10 Nov 1989	F/A-18C	CV-66	CVW-1	AB-4xx	Med, Indian Ocean
VFA-86	28 Dec 1990–18 Apr 1991	F/A-18C	CV-66	CVW-1	AB-4xx	Arabian Gulf
VFA-86	21 Aug 1991–11 Oct 1991	F/A-18C	CV-66	CVW-1	AB-4xx	NorLant
VFA-86	02 Dec 1991–06 Jun 1992	F/A-18C	CV-66	CVW-1	AB-4xx	Med
VFA-86	11 Aug 1993–05 Feb 1994	F/A-18C	CV-66	CVW-1	AB-4xx	Med
VFA-86	28 Aug 1995–24 Feb 1996	F/A-18C	CV-66	CVW-1	AB-4xx	Med, Persian Gulf
VFA-86	03 Oct 1997–03 Apr 1998	F/A-18C	CVN-73	CVW-1	AB-4xx	Med, Persian Gulf
VFA-86	02 Nov 1998–17 Dec 1998	F/A-18C	CVN-75	CVW-1	AB-4xx	Puerto Rican Operating Area
VFA-86	22 Sep 1999–19 Mar 2000	F/A-18C	CV-67	CVW-1	AB-4xx	Med, Persian Gulf
VFA-86	19 Sep 2001–27 Mar 2002	F/A-18C	CVN-71	CVW-1	AB-4xx	Med, Northern Arabian Sea
VFA-86	28 Aug 2003 29 Feb 2004	F/A-18C	CVN-65	CVW-1	AB-4xx	SoLant, Med, Persian Gulf
VFA-86	03 Jun 2004–23 Jul 2004	F/A-18C	CVN-65	CVW-1	AB-4xx	EastLant, NorLant
VFA-86	02 May 2006–18 Nov 2006	F/A-18C(N)	CVN-65	CVW-1	AB-4xx	Med, Mid-East, WestPac
VFA-86	07 Jul 2007–19 Dec 2007	F/A-18C(N)	CVN-65	CVW-1	AB-4xx	Med, Mid-East
VFA-86	31 Jul 2009–26 Mar 2010	F/A-18C(N)	CVN-68	CVW-11	NH-4xx	WestPac, north Arabian Sea

VFA-87 Golden Warriors

The Golden Warriors entered a new age on 24 October 1986 which, akin to many other Navy attack squadrons, saw the esteemed LTV A-7E Corsair II retired and the squadron designated Strike Fighter Squadron 87 with the new McDonnell Douglas F/A-18A Hornet.

After a short-term assignment within Carrier Air Wing One (CVW-1), the squadron boarded USS *Theodore Roosevelt* (CVN-71) in December 1988 joining Carrier Air Wing Eight (CVW-8), Then, on 28 December 1990, VFA-87 deployed for Operation Desert Storm, flying 629 sorties over forty-three days to free Kuwait before returning to the USA in June 1991 and transitioning to the newer F/A-18C version of the Boeing Hornet.

From March to September 1993, the Golden Warriors supported Operations Deny Flight, Provide Comfort and Southern Watch, its first deployment with the night-attack F/A-18C Hornet. USS *Theodore Roosevelt* (CVN-71) and VFA-87 returned to the Arabian Gulf in March 1995 for Southern Watch followed by combat sorties over Yugoslavia. On 30 August 1995 during Operation Deliberate Force, Golden Warrior jets were the first to hit Bosnian Serb targets.

Armed with JDAM and AIM-9L missiles, F/A-18C 164205 is seen aboard the USS *Theodore Roosevelt* during Operation Iraqi Freedom during March 2003.

During operations over Afghanistan during February 2009 VFA-87 had temporarily transitioned temporarily back to the F/A-18A+ version. Aircraft 162872 'AJ-402', call sign 'Commando 45', is seen high over the country just after sunset.

By 2011 the squadron still hadn't regained its F/A-18C mounts. Here aircraft 163107 is seen sporting not only the CO's colourful tail but also a 100th anniversary marking of the US Navy on the inboard of the starboard fin.

Participation in Operation Deliberate Guard during a summer deployment with USS *John F. Kennedy* (CV-67) followed. VFA-87 also returned to the familiar Arabian Gulf for Southern Watch before coming home in October of 1997 to NAS Cecil Field for the very last time.

VFA-87 famously participated in combat operations in two different theatres during a single deployment, operating from USS *Theodore Roosevelt* (CVN-71). While on its way to the Arabian Gulf in April 1999, a crisis broke out in the former Yugoslavian republic of Kosovo, precipitating the largest aerial bombing campaign since the Vietnam War. 'War Party' F/A-18s dropped 430,000lb of ordnance and flew 595 combat missions during Operation Allied Force. All of this assisted in the eventual NATO victory and removal of Serbian oppressors from Kosovo.

July saw VFA-87 return to the Arabian Gulf, where it flew 176 combat missions in support of Operation Southern Watch. Following this deployment, the Golden Warriors returned to NAS Oceana in Virginia Beach, Virginia, following the closure of NAS Cecil Field under the BRAC initiative. A Southern Watch support cruise followed in April 2001 aboard the 'Big E', which culminated in a great offensive against the Taliban and Al-Qaeda forces in Afghanistan that marked the birth of a war against global terrorism in Operation Enduring Freedom.

Fatigue issues among the Hornet fleet saw Strike Fighter Squadron 87 complete an unprecedented and highly successful transition from the heavily utilised, combat-proven F/A-18C back to the other older but less fatigued F/A-18A+ Hornet in late 2006. Throughout 2007 and the beginning of 2008, VFA-87 participated in a number of detachments and preparation exercises in readiness for their next deployment. This occurred with a deployment on the USS *Theodore Roosevelt* (CVN-71) in support of Operation Enduring Freedom in September, going back to NAS Oceana in April 2009.

In 2012, the Golden Warriors accepted ten F/A-18C aircraft in a straight swap once again with VFC-12 and began preparing them for an intense period of preparation and future deployment. The aircraft they were preparing had not been to sea or dropped air-to-surface ordnance in upwards of five years. The end of 2012 brought the beginning of another period of training, and in 2013 VFA-87 completed six detachments around the country in preparation for deployment. In February 2014, VFA-87 again embarked aboard CVN-77 as part of the Fifth Fleet area of responsibility in support of National Command Authority tasking.

The squadron was to then switch to the more capable F/A-18E Super Hornet during fiscal year 2016.

VFA-87	04 Jan 1987–03 Feb 1987	F/A-18A	CVN-71	CVW-1	AC-4xx	Shakedown (Guantanamo Bay)
VFA-87	25 Aug 1988–11 Oct 1988	F/A-18A	CVN-71	CVW-8	AJ-4xx	NorLant
VFA-87	30 Dec 1988–30 Jun 1989	F/A-18A	CVN-71	CVW-8	AJ-4xx	Med
VFA-87	19 Jan 1990–23 Feb 1990	F/A-18A	CVN-72	CVW-8	AJ-4xx	Shakedown (SoLant)
VFA-87	28 Dec 1990–28 Jun 1991	F/A-18A	CVN-71	CVW-8	AJ-4xx	Desert Shield & Desert Storm
VFA-87	11 Mar 1993–08 Sep 1993	F/A-18C(N)	CVN-71	CVW-8	AJ-4xx	Med
VFA-87	22 Mar 1995–22 Sep 1995	F/A-18C(N)	CVN-71	CVW-8	AJ-4xx	Med
VFA-87	01 Mar 1996–31 Mar 1996	F/A-18C(N)	CVN-74	CVW-8	AJ-4xx	Shakedown (SoLant)
VFA-87	17 May 1996–28 Jun 1996	F/A-18C(N)	CV-67	CVW-8	AJ-4xx	NorLant
VFA-87	29 Apr 1997–28 Oct 1997	F/A-18C(N)	CV-67	CVW-8	AJ-4xx	Med, Persian Gulf
VFA-87	26 Mar 1999–22 Sep 1999	F/A-18C(N)	CVN-71	CVW-8	AJ-4xx	Med, Persian Gulf
VFA-87	25 Apr 2001–10 Nov 2001	F/A-18C(N)	CVN-65	CVW-8	AJ-4xx	Med, Persian Gulf
VFA-87	06 Jan 2003–29 May 2003	F/A-18C(N)	CVN-71	CVW-8	AJ-4xx	Caribbean, Med
VFA-87	01 Sep 2005–11 Mar 2006	F/A-18C(N)	CVN-71	CVW-8	AJ-4xx	Med, Persian Gulf
VFA-87	08 Sep 2008–18 Apr 2009	F/A-18C(N)	CVN-71	CVW-8	AJ-4xx	Med, CENTCOM AOR
VFA-87	11 May 2011–10 Dec 2011	F/A-18A+	CVN-77	CVW-8	AJ-4xx	Med, north Arabian Sea, Persian Gulf
VFA-87	15 Feb 2014–15 Nov 2014	F/A-18C(N)	CVN-77	CVW-8	AJ-4xx	Med, north Arabian Sea, Persian Gulf

VFA-94 Shrikes

In June 1990, the squadron received its first F/A-18C(N) Hornet and was re-designated Strike Fighter Squadron 94 (VFA-94) on 1 January 1991. The following May it deployed to the Persian Gulf aboard USS *Abraham Lincoln* (CVN-72). This was in support of United Nations sanctions after the Iraq war. The squadron also played an active role in the disaster relief caused by the eruption of Mount Pinatubo in the Philippines, including the evacuation of people made homeless by the volcano in Operation Fiery Vigil.

June 1993 saw the Shrikes fly missions over southern Iraq in support of Operation Southern Watch and later in October 1993 they departed the Persian Gulf for Somalia to provide force protection for US and other UN forces working together to provide humanitarian relief.The next deployment, which began on 10 October 1996, was to the Persian Gulf aboard USS *Kitty Hawk* (CV-63) in support of Operation Southern Watch before cross-decking with the remainder of the wing, CVW-11, in May 1997 aboard USS *Carl Vinson*. The next twelve months were to see three deployments to the Persian Gulf in support of Operation Desert Fox, an EastPac cruise and RIMPAC Exercise.

After the 11 September attacks, the squadron led the first missions of Operation Enduring Freedom on 7 October 2001. Immediately following this, VFA-94 and Air Wing Eleven took part in precision strikes against key Taliban locations in Afghanistan. The squadron accumulated 664 combat sorties and expended 550,000lb of ordnance on targets in Afghanistan, before returning on 19 January 2002.

The next eight-month combat deployment began in March 2003 aboard USS *Nimitz* with CVW-11 in support of Operation Iraqi Freedom, with a second beginning on 7 May 2005, when the squadron again deployed aboard the *Nimitz* to the Persian Gulf in support of Operation Iraqi Freedom.

In October 2006, the squadron underwent preparations for transition to an expeditionary role, including the completion of SFARP. In January 2007, VFA-94 joined its sister squadron, VFA-97, as one of only two nacy F/A-18 squadrons to deploy overseas as part of the Unit Deployment Program (UDP). VFA-94 departed NAS Lemoore on its first expeditionary deployment to MCAS Iwakuni, Japan. The squadron had to cover nearly 7,000 nautical miles (13,000km) of open ocean, totalling almost eighteen flight hours per jet. During the deployment, VFA-94 participated in Exercise Foal Eagle, which had the distinction of being the first US Navy expeditionary deployment to Korea, and Exercise Cobra Gold in Thailand.

Between 2005 and 2014 VFA-94 was not assigned to a particular air wing and following successful completion of the Strike Fighter Readiness programme it joined VFA-97 in the Unit Deployment Program (UDP) to MCAS Iwakuni, Japan. 164262 is seen here between such deployments at NAS Lemoore minus any CVW markings.

In July 2008, VFA-94 completed its second UDP deployment, during which time it participated in Exercises Talon Vision at Clark Air Base, Philippines; Wolmi-Do Fury at Kadena Air Base, Okinawa and Exercise Lava Viper at Hickam AFB, Hawaii. The squadron returned home in February 2009 after finishing their seven-month deployment.

In January 2012, after coming back from further UDP duties in Japan, the squadron returned to NAS Lemoore, but it seems they were not assigned to one of the west coast air wings until around August 2012. This appears to have been a result of stepping in to take the place of sister squadron VFA-25 within CVW-17. VFA-25 was later returned to its former slot within CVW-17, thus replacing VFA-94.

VFA-94	12 Mar 1991–16 Apr 1991	F/A-18C(N)	CVN-72	CVW-11	NH-4xx	EastPac
VFA-94	28 May 1991–25 Nov 1991	F/A-18C(N)	CVN-72	CVW-11	NH-4xx	WestPac, Persian Gulf
VFA-94	14 Oct 1992–13 Nov 1992	F/A-18C(N)	CVN-72	CVW-11	NH-4xx	EastPac
VFA-94	13 Jan 1993–12 Feb 1993	F/A-18C(N)	CVN-72	CVW-11	NH-4xx	EastPac
VFA-94	02 Mar 1993–28 Mar 1993	F/A-18C(N)	CVN-72	CVW-11	NH-4xx	EastPac
VFA-94	15 Jun 1993–15 Dec1993	F/A-18C(N)	CVN-72	CVW-11	NH-4xx	WestPac, Persian Gulf, Indian Ocean
VFA-94	11Apr1995–09 Oct 1995	F/A-18C(N)	CVN-72	CVW-11	NH-4xx	WestPac, Persian Gulf
VFA-94	11 Oct 1996–11 Apr 1997	F/A-18C(N)	CV-63	CVW-11	NH-4xx	WestPac, Persian Gulf
VFA-94	14 Mar 1998–09 Apr 1998	F/A-18C(N)	CVN-70	CVW-11	NH-4xx	EastPac
VFA-94	22 Jun 1998–11 Aug 1998	F/A-18C(N)	CVN-70	CVW-11	NH-4xx	JTFEX/FLEETEX, RIMPAC 98
VFA-94	06 Nov 1998–06 May 1999	F/A-18C(N)	CVN-70	CVW-11	NH-4xx	WestPac, Persian Gulf
VFA-94	23 Jul 2001–23 Jan 2002	F/A-18C(N)	CVN-70	CVW-11	NH-4xx	WestPac, Persian Gulf
VFA-94	03 Mar 2003–05 Nov 2003	F/A-18C(N)	CVN-68	CVW-11	NH-4xx	WestPac, Persian Gulf
VFA-94	07 May 2005–08 Nov 2005	F/A-18C(N)	CVN-68	CVW-11	NH-4xx	WestPac, Persian Gulf
VFA-94	22 Aug 2014–04 Jun 2015	F/A-18C(N)	CVN-70	CVW-17	NA-4xx	WestPac, CENTCOM AOR

VFA-97 Warhawks

In 1990 the squadron completed its fifteenth WestPac deployment on the LTV A-7E Corsair II, which was also their final one. On 24 January 1991 the squadron was re-designated Strike Fighter Squadron 97 (VFA-97) and began transition to the F/A-18A Hornet; after this, in the summer, the squadron went to NAS Norfolk and embarked on USS *Kitty Hawk* (CV-63) for her passage around Cape Horn.

In 1992 the squadron deployed to the WestPac in support of Operation Southern Watch over Iraq as well as Operation Restore Hope in Somalia, before heading back to NAS Lemoore in May 1993. The squadron's next cruise began in June 1994 (its seventeenth overall and third with the Hornet). This was its final cruise as a member of CVW-15 and in October 1995 the squadron joined Carrier Air Wing Eleven (CVW-11), although once again aboard *Kitty Hawk*, where it deployed in support of Operation Southern Watch.

In December 1997 Carrier Air Wing Eleven and the squadron were assigned to USS *Carl Vinson*, CVN-70, deploying on 10 November 1998 for WestPac 98–99, its nineteenth deployment. The squadron supported Operation Desert Fox and Operation Southern Watch in flying combat missions.

July 2001 saw the squadron aboard the USS *Carl Vinson* and following the 11 September attacks the carrier and her air wing hurried to the north Arabian Sea and carried out sustained combat operations in support of Operation Enduring Freedom. The squadron completed more than 3,000 flight hours, 1,340 sorties, and had a 99 per cent combat sortie completion rate, delivering more than 453,000lb of ordnance in a ten-week period.

In 2003, the squadron, still attached to CVW-11, embarked on USS *Nimitz* (CVN-68) and flew hundreds of combat sorties in support of Operation Iraqi Freedom. This was to be the last time the squadron's F/A-18As would be used in combat. After return from deployment, the squadron transitioned from the F/A-18A to the F/A-18C and began preparation for its upcoming Unit Deployment Program (UDP) deployment.

In 2004, VFA-97 was assigned to Marine Aircraft Group Twelve. They were the first Navy squadron to participate in the UDP, deploying to MCAS Iwakuni, Japan. The squadron deployed to MCAS Iwakuni in 2006 for its second UDP deployment, and again in

Seen early in its career as a Hornet user, this VFA-97 F/A-18A was assigned to CVW-15 aboard the USS *Kittyhawk*. The squadron undertook three such cruises with this wing. (Photo US Navy via Geoff Rhodes)

This VFA-97 jet, 163493, is seen at NAS Lemoore between UDP deployments and therefore lacking any air wing identification.

Seen in 2011 at the NAS North Island NAD (Naval Air Depot), 164013, awaits its turn to enter a major rebuild wearing marks from its 2003 WestPac cruise. It later returned to active strength with the NASWC.

January 2008 for its third UDP which was actually their twenty-fourth deployment all in all. VFA-97 re-joined CVW-11 and *Nimitz* on return from Iwakuni and began a compressed turnaround for deployment on board the carrier. In February 2009, VFA-97 embarked on the *Nimitz* in the night interdiction role on the F/A-18C(N), and with that its five-year departure from a carrier air wing was over.

VFA-97 later transferred to Carrier Air Wing Nine (CVW-9) for two cruises aboard CVN-74, the latter ending in May 2013 and following which the squadron started transitioning to the F/A-18E Super Hornet.

VFA-97	18 Oct 1991–11 Dec 1991	F/A-18A	CV-63	CVW-15	NL-3xx	Norfolk to San Diego
VFA-97	03 Nov 1992–03 May 1993	F/A-18A	CV-63	CVW-15	NL-3xx	WestPac, Persian Gulf
VFA-97	24 Jun 1994–22 Dec 1994	F/A-18A	CV-63	CVW-15	NL-3xx	WestPac, Indian Ocean
VFA-97	11 Oct 1996–11 Apr 1997	F/A-18A	CV-63	CVW-11	NH-2xx	WestPac, Persian Gulf
VFA-97	14 Mar 1998–09 Apr 1998	F/A-18A	CVN-70	CVW-11	NH-2xx	EastPac
VFA-97	22 Jun 1998–11 Aug 1998	F/A-18A	CVN-70	CVW-11	NH-2xx	JTFEX/FLEETEX, RIMPAC 98
VFA-97	06 Nov 1998–06 May 1999	F/A-18A	CVN-70	CVW-11	NH-2xx	WestPac, Persian Gulf
VFA-97	23 Jul 2001–23 Jan 2002	F/A-18A	CVN-70	CVW-11	NH-2xx	WestPac, Persian Gulf
VFA-97	03 Mar 2003–05 Nov 2003	F/A-18A	CVN-68	CVW-11	NH-3xx	WestPac, Persian Gulf
VFA-97	Sep 2004–Mar 2005	F/A-18C	MAG-12/MCAS			Iwakuni, Japan
VFA-97	Mar 2006–Sep 2006	F/A-18C	MAG-12/MCAS			Iwakuni, Japan
VFA-97	Jan 2008–Jul 2008	F/A-18C	MAG-12/MCAS			Iwakuni, Japan
VFA-97	31 Jul 2009–26 Mar 2010	F/A-18C(N)	CVN-68	CVW-11	NH-3xx	WestPac, north Arabian Sea
VFA-97	25 Jul 2011–02 Mar 2012	F/A-18C(N)	CVN-74	CVW-9	NG-3xx	WestPac, north Arabian Sea, Persian Gulf
VFA-97	27 Aug 2012–03 May 2013	F/A-18C(N)	CVN-74	CVW-9	NG-3xx	WestPac, north Arabian Sea

VFA-105 Gunslingers

On 17 December 1990, VA-105 was re-designated a Strike Fighter Squadron, as the unit transitioned to the F/A-18C(N); following this, it was assigned to CVW-3 aboard the USS *John F. Kennedy* (CV-67) in the following September. VFA-105, as it was now named, made its first Hornet cruise a year later in October 1992 aboard the 'Big John'. The squadron came back from the Mediterranean in April 1993, and began training for another Mediterranean deployment, this time aboard CVN-69, the USS *Dwight D. Eisenhower*, in October 1994. The Gunslingers cross-decked for a deployment aboard the USS *Theodore Roosevelt* (CVN-71) in October 1996. In 1998 VFA-105 and the rest of CVW-3 were deployed in support of Operation Southern Watch and Operation Desert Fox on board USS *Enterprise*.

164244 'AC-404' seen at its home shore base of NAS Oceana.

After a somewhat nomadic existence, on 28 November 2000 the squadron and remainder of Air Wing Three were embarked on the maiden deployment of USS *Harry S. Truman* (CVN-75) to the Mediterranean and Persian Gulf in support of Operation Southern Watch. The Gunslingers deployed in support of Operation Iraqi Freedom aboard the *Truman* in 2003, operating from the Mediterranean Sea, the first of three such deployments aboard the carrier.

VFA-105 then became the first east coast squadron to transition to the F/A-18E Super Hornet, in July 2006. In March 2007, VFA-105 made history in becoming the first operational Navy strike fighter squadron to be commanded by a woman after Cdr Sara Joyner took command.

VFA-105	07 Oct 1992–07 Apr 1993	F/A-18C(N)	CV-67	CVW-3	AC-4xx	Med
VFA-105	20 Oct 1994–14 Apr 1995	F/A-18C(N)	CVN-69	CVW-3	AC-4xx	Med
VFA-105	25 Nov 1996–22 May 1997	F/A-18C(N)	CVN-71	CVW-3	AC-4xx	Med
VFA-105	06 Nov 1998–06 May 1999	F/A-18C(N)	CVN-65	CVW-3	AC-4xx	Med, Adriatic Sea, Persian Gulf
VFA-105	28 Nov 2000–23 May 2001	F/A-18C(N)	CVN-75	CVW-3	AC-4xx	Med, Persian Gulf
VFA-105	05 Dec 2002–23 May 2003	F/A-18C(N)	CVN-75	CVW-3	AC-4xx	Med
VFA-105	02 Jun 2004–25 Jul 2004	F/A-18C(N)	CVN-75	CVW-3	AC-4xx	Lant, Med
VFA-105	13 Oct 2004–18 Apr 2005	F/A-18C(N)	CVN-75	CVW-3	AC-4xx	Med, Persian Gulf

VFA-106 Gladiators

VFA-106 was activated at NAS Cecil Field, Florida, on 27 April 1984, flying the F/A-18A/B Hornet as part of the East Coast Fleet Replacement Air Group (RAG).

The first Replacement Pilot Class began training on the F/A-18 Hornet on 7 October 1985. According to the US Navy mission statement, as the East Coast Fleet Replacement Squadron, the squadron's mission is to train Navy and Marine Corps F/A-18 replacement pilots and weapon systems officers (WSOs) to support fleet commitments. Every six weeks, a class of eight to twelve newly winged Navy and Marine Corps pilots and Naval flight officers begins the nine-month training course in which they learn the basics of air-to-air and air-to-ground missions, culminating in day/night carrier qualification and subsequent assignment to fleet Hornet squadrons.

In October and December 1987, VFA-106 received its first C and D Hornets. In mid 1999, after the BRAC-mandated closure of NAS Cecil Field, VFA-106 relocated to NAS Oceana. In 2004, VFA-106 received its first F/A-18E/F Super Hornets and continues to train crews on all four variants.

Unlike the US Marine Corps, the Navy only employed the two-seat F/A-18D Hornet in the training role and as such these were generally only assigned to the Replacement Air Squadrons. Here 163454 is seen early in its career. The jet was lost in an accident on 4 March 1992. (Photo Robbie Shaw)

VFA-113 Stingers

On 25 March 1983 the Strike Fighter Squadron era began as VA-113 was re-designated VFA-113 and traded its steadfast LTV A-7E Corsair II for the new F/A-18A Hornet. As one of the longest-serving fleet squadrons the Stingers completed the transition to the new multi-role fighter and on 14 December 1983 became the first fleet operational, combat ready Strike Fighter Squadron in the US Navy.

Assigned to CVW-14, the squadron embarked for the first carrier deployment of the F/A-18A on board the USS *Constellation* (CV-64) in February 1985. The squadron was to undertake five such cruises aboard the *Connie* before being slated to convert to the more advanced F/A-18C variant, which took place in 1989.

After many deployments flying the Hornet throughout the late 1980s, the Stingers again sailed to the Western Pacific in June 1990, this time on board USS *Independence* (CV-62). The Persian Gulf crisis was big news story, and the Stingers were combat ready and available.

Seen at NAS Lemoore just prior to its final cruise with the F/A-18C(N), 164658 sports the 'NA' code of CVW 70 and seems to have borrowed a centre line tank from VFA 146, who were at the time in transition to the Super Hornet.

Following the invasion of Kuwait by Iraqi forces in August 1990, USS *Independence*, as part of Battle Group Delta, was the first aircraft carrier on station. VFA-113 undertook missions flown over the northern Persian Gulf, and assisted the US Navy in keeping Iraqi hostility at bay until other US and foreign forces could arrive on the scene.

The Stingers made additional deployments to the Western Pacific and to the Persian Gulf, participating in Operation Southern Watch and Operation Desert Strike throughout the mid 1990s while stationed on board USS *Carl Vinson* (CVN-70).

In March 1997, history was made by Lt Keith 'Gunny' Henderson and the Stingers as he broke the 100,000 mishap-free flight hour mark while flying aircraft 301. This was unprecedented at the time, making the Stingers the safest recorded tactical aircraft squadron in aviation history. A year later, in 1998 the squadron marked the fiftieth anniversary of its commissioning.

Between 1998 and 2006, VFA-113 made numerous deployments from both the USS *Abraham Lincoln* (CVN-72) and USS *John C. Stennis* (CVN-74), contributing to both Operation Iraqi Freedom and Operation Enduring Freedom after the 11 September attacks.

In 2006 VFA-113 moved along with Carrier Air Wing Fourteen (CVW-14) to deploy from the US's newest operational carrier, USS *Ronald Reagan* (CVN-76), taking part in the *Reagan*'s maiden deployment from January through July 2006 operating throughout the Western Pacific and the Arabian Gulf and flying missions in support of Operation Iraqi Freedom.

Through 2007 and 2008 VFA-113 undertook two further deployments aboard CVN-76, including missions in support of Coalition forces on the ground in Afghanistan for seven weeks, before returning home in November 2008. There, just months later, the squadron exceeded 150,000 hours of class-A mishap-free flight time. Lt Andy 'Shortburn' Toll was in aircraft 302 when this was achieved, during a day carrier qualification flight while flying from USS *Ronald Reagan* off the coast of Southern California. This occurred nearly thirty-five years after the Stingers lost an A-7E Corsair II (156825) after an engine failure on 12 May 1974.

A further deployment aboard CVN-76 was undertaken in 2009 before the squadron was re-assigned to CVW-17 and the USS *Carl Vinson*. The Stingers' final cruise with the first generation jets started in August 2014, when the squadron embarked on its final combat deployment with the F/A-18C and Carrier Air Wing Seventeen (CVW-17) aboard USS *Carl Vinson* (CVN-70). The Stingers flew 367 combat missions for Operation Inherent Resolve. The squadron continued to fly the F/A-18C until 17 February 2016, when it undertook its last mission on the legacy variant before the transition to the Super Hornet.

VFA-113	20 Feb 1985–24 Aug 1985	F/A-18A	CV-64	CVW-14	NK-3xx	WestPac, Indian Ocean
VFA-113	04 Sep 1986–20 Oct 1986	F/A-18A	CV-64	CVW-14	NK-3xx	NorPac
VFA-113	11 Apr 1987–13 Oct 1987	F/A-18A	CV-64	CVW-14	NK-3xx	WestPac, Indian Ocean, Arabian Gulf
VFA-113	01 Dec 1988 – 01 Jun 1989	F/A-18A	CV-64	CVW-14	NK-3xx	WestPac, Indian Ocean
VFA-113	16 Sep 1989–19 Oct 1989	F/A-18A	CV-64	CVW-14	NK-3xx	NorPac
VFA-113	27 Apr 1990–20 May 1990	F/A-18A	CV-62	CVW-14	NK-3xx	RIMPAC
VFA-113	23 Jun 1990–20 Dec 1990	F/A-18A	CV-62	CVW-14	NK-3xx	WestPac, Persian Gulf
VFA-113	05 Aug 1991–22 Aug 1991	F/A-18C	CV-62	CVW-14	NK-3xx	west coast to Hawaii
VFA-113	28 Aug 1991–14 Sep 1991	F/A-18A	CV-41	CVW-14	NK-3xx	Hawaii to west coast
VFA-113	17 Feb 1994–17 Aug 1994	F/A-18C(N)	CVN-70	CVW-14	NK-3xx	WestPac, Persian Gulf
VFA-113	21 Aug 1995–13 Sep 1995	F/A-18C(N)	CVN-70	CVW-14	NK-3xx	Hawaii Operating Area
VFA-113	14 May 1996–14 Nov 1996	F/A-18C(N)	CVN-70	CVW-14	NK-3xx	WestPac, Persian Gulf
VFA-113	04 Apr 1997–28 Apr 1997	F/A-18C(N)	CVN-72	CVW-14	NK-3xx	EastPac
VFA-113	22 Nov 1997–20 Dec 1997	F/A-18C(N)	CVN-72	CVW-14	NK-3xx	EastPac
VFA-113	11 Jun 1998–11 Dec 1998	F/A-18C(N)	CVN-72	CVW-14	NK-3xx	WestPac, Persian Gulf
VFA-113	12 May 2000–01 Jul 2000	F/A-18C(N)	CVN-72	CVW-14	NK-3xx	RIMPAC 2000
VFA-113	17 Aug 2000–12 Feb 2001	F/A-18C(N)	CVN-72	CVW-14	NK-3xx	WestPac, Persian Gulf
VFA-113	14 Apr 2002–14 May 2002	F/A-18C(N)	CVN-72	CVW-14	NK-3xx	NorPac, JTFEX
VFA-113	20 Jul 2002–06 May 2003	F/A-18C(N)	CVN-72	CVW-14	NK-3xx	WestPac, Persian Gulf
VFA-113	24 May 2004–01 Nov 2004	F/A-18C(N)	CVN-74	CVW-14	NK-3xx	RIMPAC 04, WestPac
VFA-113	04 Jan 2006–06 Jul 2006	F/A-18C(N)	CVN-76	CVW-14	NK-3xx	WestPac, Persian Gulf
VFA-113	27 Jan 2007–20 Apr 2007	F/A-18C(N)	CVN-76	CVW-14	NK-3xx	WestPac
VFA-113	19 May 2008–25 Nov 2008	F/A-18C(N)	CVN-76	CVW-14	NK-3xx	WestPac, north Arabian Sea
VFA-113	28 May 2009–21 Oct 2009	F/A-18C(N)	CVN-76	CVW-14	NK-3xx	WestPac, north Arabian Sea
VFA-113	30 Nov 2010–15 Jun 2011	F/A-18C(N)	CVN-70	CVW-17	AA-3xx	COMPTUEX, WestPac, north Arabian Sea
VFA-113	30 Nov 2011–23 May 2012	F/A-18C(N)	CVN-70	CVW-17	AA-3xx	WestPac, CENTCOM AOR
VFA-113	22 Aug 2014–04 Jun 2015	F/A-18C(N)	CVN-70	CVW-17	NA-3xx	WestPac, CENTCOM AOR

VFA-115 Eagles

VFA-115 did not become operational on the Boeing F/A-18 Hornet until 1996. Attack Squadron VA-115, with its Grumman A-6E Intruder aircraft as part of Carrier Air Wing Five (CVW-5), flew off the USS *Midway* (CV-41) and back to the US on 24 March 1986, and the carrier was then modified to allow the F/A-18 Hornet to operate from her deck. In the following October the squadron, one of the last to operate the Intruder, TransPac'd with the rest of the 'new' CVW-5 to NAF Atsugi in Japan and the newly refurbished CV-41, on whom the unit was to operate for a further five years.

In August 1991 the USS *Midway* left Yokosuka, Japan, for the last time to be decommissioned and eventually become a floating museum at its US west coast home port of San Diego. In the period between 22 and 27 August CVW-5 transferred from the *Midway* to the USS *Independence* at Pearl Harbor in what is known as 'The Great Carrier Air Wing SwapEx'.

VA-115, with its Intruder aircraft, and the *Independence*, CV-62, remained forward deployed for a further five years. In 1996 the last Intruder squadrons were decommissioned as the A-6E was retired. The lone squadron that remained active at that time was VA-115 but it too began transition to the F/A-18C Hornet after the completion of RIMPAC 96 in June that year, which saw the squadron relocate to NAS Lemoore.

In October 1996 the squadron was re-designated VFA-115. The Eagles' new carrier air wing also transitioned to CVW-14 aboard the USS *Abraham Lincoln* (CVN-72); it was to undertake five cruises with the legacy Hornet.

Its third cruise aboard CVN-72 in June 1998 the squadron deployed to the Arabian Gulf in support of Operation Southern Watch but, in spite of being one of the last Navy Intruder squadrons to transition to the Hornet, VFA-115 was chosen to be the first Navy fleet squadron to receive the new F/A-18E Super Hornet and became combat ready on the type in 2002.

VFA-115	04 Apr 1997–28 Apr 1997	F/A-18C	CVN-72	CVW-14	NK-2xx	EastPac
VFA-115	22 Nov 1997–20 Dec 1997	F/A-18C	CVN-72	CVW-14	NK-2xx	EastPac
VFA-115	11 Jun 1998–11 Dec 1998	F/A-18C	CVN-72	CVW-14	NK-2xx	WestPac, Persian Gulf
VFA-115	12 May 2000–01 Jul 2000	F/A-18C	CVN-72	CVW-14	NK-2xx	RIMPAC 2000
VFA-115	17 Aug 2000–12 Feb 2001	F/A-18C	CVN-72	CVW-14	NK-2xx	WestPac

VFA-122 Flying Eagles

In January 1999 a new Flying Eagles squadron was established as Strike Fighter Squadron 122 (VFA-122), and it later became the first unit to operate the F/A-18E/F.

Nearly eleven years later, on 1 October 2010, VFA-122 was merged with VFA-125 (by then the legacy F/A-18 Hornet FRS and also stationed at NAS Lemoore). The move was carried out with the aim of cutting administrative costs and making production more efficient in advance of the phasing out of the legacy F/A-18 Hornet by the Super Hornet and F-35 Lightning II, scheduled across the coming years. The merged squadron kept its Flying Eagles insignia but the Rough Raiders of VFA-125 were put on hold until 2017, when it was re-established as an F-35 training squadron.

As the west coast Hornet and Super Hornet Fleet Replacement Squadron, the unit's mission has been training Navy and Marine Corps F/A-18A/B/C/D/E/F replacement pilots and weapon systems officers (WSOs) to assist with fleet commitments.

Legacy Hornet training had all but ceased at NAS Lemoore by 2017 as delivery of the new Lockheed Martin F-35C had begun. Remaining early derivative training passing solely into the hands of the East Coast RAG, VFA-106, at NAS Oceana.

When VFA-125 stood down in preparation to transition to the F-35C, all west coast Hornet training became the responsibility of VFA-122. However, the use of the legacy Hornet was short-lived and 164011 was one of the last in use. (Photo Tony Osborne)

Seen just prior to the squadron decommissioning, F/A-18C 163726 sports the commander's all-black tail and enhanced markings.

In clean configuration, F/A-18C 164034 makes a high-speed pass at Edwards AFB during the 2006 open house.

VFA-125 Rough Raiders

VFA-125 Rough Raiders became the Navy's first F/A-18 squadron when it was established on 13 November 1980 at NAS Lemoore in California VFA-125 received its first aircraft in April 1981, and by March 1985 had accumulated more than 30,000 accident-free flight hours.

On 1 October 2010, after thirty years in the Hornet RAG role, VFA-125 was disestablished as an F/A-18 Fleet Replacement Squadron and its aircraft and personnel were incorporated into VFA-122. It was planned that VFA-125 would be re-established as an F-35 Lightning II training squadron in the future and to that effect received its first F-35C during the first quarter of 2017.

VFA-127 Desert Bogeys

VFA-127 was one of the more short-lived US Navy Hornet squadrons, operating the type for less than a decade. Before being established as VA-127 in June 1962, the unit operated as VA-126 Detachment Alpha. Its mission was to deliver basic and refresher all-weather jet instrument and transition training for west coast pilots. Upon being re-designated VA-127, its primary mission became to provide advanced all-weather jet instrument training for fleet replacement pilots and refresher training for light attack pilots, with a secondary mission of jet transition and refresher training. With the arrival of the Boeing F/A-18 Hornet in 1987 at NAS Fallon it was able to enhance this task and it was re-designated VFA-127 accordingly on 1 March 1987. Alas, its tenure in the Hornet world was relatively short lived and the squadron was disestablished on 23 March 1996, with some of its former mounts being sold on the Spanish Air Force to re-equip Ala 46 at Gando.

Single-seat F/A-18A 162899 also sporting the VFA-127 unit's brown aggressor colour scheme when seen at NAS Fallon in 1995.

This jet is 165213 'AG-402' of VFA-131 seen off the US eastern seaboard on 12 March 1996.

VFA-131 Wildcats

The Wildcats took on their first F/A-18A Hornet in May 1984 having been established as VFA-131 at NAS Lemoore the previous October. They moved to NAS Cecil Field, Florida, in January 1985, where they became 'AIRLANT'S' (Naval Air Forces Atlantic) first F/A-18 squadron and were to become one of the US Navy's longest-serving squadrons with the legacy Hornet.

As part of Carrier Air Wing Thirteen (CVW-13), the squadron undertook its first cruise with the Hornet when it deployed to the Mediterranean Sea in October 1985, from where it participated in air strikes against Libya. The unit was on board USS *Coral Sea* (CV-43).

The Wildcats completed their second Mediterranean deployment on the *Coral Sea* in March 1988 and then in August VFA-131 embarked in USS *Independence* (CV-62) for a two-month transit via Cape Horn from Norfolk to San Diego. Here the squadron joined Carrier Air Wing Seven (CVW-7) and deployed to the Mediterranean Sea on board USS *Dwight D. Eisenhower* (CVN-69) in March 1990. It was among those units from the United States Armed Forces to respond to the Iraq invasion of Kuwait as part of Operation Desert Shield.

In September 1990, after six years of the 'Alpha' model of the Hornet, the squadron transitioned to F/A-18C Lot XIII Night Strike Hornets. This was a variant the unit was to take to war aboard CVN-69 in support of Operation Desert Storm. This deployment ended with operations above the Arctic Circle supporting Exercise Teamwork 92.

In May 1994 the Navy's newest nuclear-powered aircraft carrier, USS *George Washington* (CVN-73), was deployed for the first time and VFA-131 was involved, flying sorties in support of Operation Deny Flight over Bosnia-Herzegovina and Operation Southern Watch over southern Iraq. In October 1994, prompted by Iraqi hostility, the squadron participated in Operation Vigilant Warrior. Further deployments in support of Operations Decisive Endeavour and Southern Watch followed.

February 1998 saw VFA-131 undertake the maiden around the world deployment of USS *John C. Stennis* (CVN-74), which included supporting Operation Southern Watch in Iraq. After they returned in December 1998, the Wildcats moved from NAS Cecil Field, Florida, to NAS Oceana in Virginia Beach, Virginia, following the BRAC decision to close the former base.

February 2000 saw the air wing re-join the USS *Dwight D. Eisenhower* and a year later, on 11 September 2001, saw VFA-131 under way on board USS *John F. Kennedy* (CV-67). Fast to respond, armed Hornets were conducting air patrols over Washington DC and New York City in support of Operation Noble Eagle within hours of the terrorist attacks being launched. VFA-131 remained an integral part of Carrier Air Wing Seven (CVW-7) until it transferred to CVW-3 in 2016. In the intervening period the squadron undertook seven carrier deployments, six as part of the Sixth Fleet in the Mediterranean. The seventh, between April and May 2008, saw the squadron temporarily join the USS *George Washington* as part of CVW-17 as it transferred from the Atlantic to Pacific fleets.

In June 2016 the squadron was re-assigned to CVW-3 joining the *Ike*, CVN-69, as the current fleet readiness carrier for a preparation cruise at the beginning of 2017. It transitioned to the F/A-18E Super Hornet upon return from the cruise.

VFA-131	01 Oct 1985–19 May 1986	F/A-18A	CV-43	CVW-13	AK-1xx	Med
VFA-131	29 Sep 1987–28 Mar 1988	F/A-18A	CV-43	CVW-13	AK-1xx	Med
VFA-131	17 Apr 1989–20 May 1989	F/A-18A	CVN-69	CVW-7	AG-4xx	WestLant
VFA-131	06 Oct 1989–02 Nov 1989	F/A-18A	CVN-69	CVW-7	AG-4xx	WestLant
VFA-131	25 Jan 1990–10 Feb 1990	F/A-18A	CVN-69	CVW-7	AG-4xx	WestLant
VFA-131	08 Mar 1990–12 Sep 1990	F/A-18A	CVN-69	CVW-7	AG-4xx	Med, Persian Gulf
VFA-131	26 Sep 1991–02 Apr 1992	F/A-18C(N)	CVN-69	CVW-7	AG-4xx	Med
VFA-131	20 May 1994–18 Nov 1994	F/A-18C(N)	CVN-73	CVW-7	AG-4xx	Med
VFA-131	26 Jan 1996–23 Jul 1996	F/A-18C(N)	CVN-73	CVW-7	AG-4xx	Med
VFA-131	26 Feb 1998–26 Aug 1998	F/A-18C(N)	CVN-74	CVW-7	AG-4xx	World Cruise, Persian Gulf
VFA-131	18 Feb 2000–18 Aug 2000	F/A-18C(N)	CVN-69	CVW-7	AG-4xx	Med, Persian Gulf
VFA-131	07 Feb 2002–17 Aug 2002	F/A-18C(N)	CV-67	CVW-7	AG-4xx	Med, north Arabian Sea
VFA-131	20 Jan 2004–26 Jul 2004	F/A-18C(N)	CVN-73	CVW-7	AG-4xx	Med, Persian Gulf
VFA-131	03 Oct 2006–23 May 2007	F/A-18C(N)	CVN-69	CVW-7	AG-4xx	Med, Persian Gulf
VFA-131	07 Apr 2008–27 May 2008	F/A-18C(N)	CVN-73	CVW-17	AG-4xx	Norfolk to San Diego
VFA-131	21 Feb 2009–30 Jul 2009	F/A-18C(N)	CVN-69	CVW-7	AG-4xx	Med, north Arabian Sea
VFA-131	02 Jan 2010–28 Jul 2010	F/A-18C(N)	CVN-69	CVW-7	AG-4xx	Med, north Arabian Sea
VFA-131	20 Jun 2012–19 Dec 2012	F/A-18C(N)	CVN-69	CVW-7	AG-4xx	Med, north Arabian Sea, Persian Gulf
VFA-131	21 Feb 2013–03 Jul 2013	F/A-18C(N)	CVN-69	CVW-7	AG-4xx	Med, north Arabian Sea
VFA-131	01 Jun 2016–30 Dec 2016	F/A-18C(N)	CVN-69	CVW-3	AC-3xx	5th and 6th Fleet AOR
VFA-131	Jan 2017– Jul 2017	F/A-18C(N)	CVN-69	CVW-3	AC-3xx	SUSTEX

F/A-18A 162451 of VFA-132 seen at NAF Kadena while on a UDP deployment as part of MAG-15 at MCAS Iwakuni during 1987. The squadron was decommissioned in June 1992 and its jets passed to the US Marine Corps.

VFA-132 Privateers

VFA-132 was established on 3 January 1984 as the first squadron to be assigned the VFA designation. It operated the F/A-18A Hornet from NAS Lemoore, California, prior to relocating to NAS Cecil Field in February 1985. The unit's maiden deployment was aboard the USS *Coral Sea* (CV-43) and it took part in combat operations during Operations Prairie Fire and El Dorado Canyon against Libya.

During Freedom of Navigation exercises in the Gulf of Sidra in March 1986, the squadron's aircraft flew combat air patrols supporting the exercise through 24–25 March after the Libyan firing of an SA-5 missile against an US aircraft operating in international waters on 24 March.

The first use of the F/A-18 in combat came when VFA-132 aircraft, along with other units of Carrier Air Wing Thirteen (CVW-13) and A-7E Corsairs from CVW-1, provided air-to-surface AGM-45 Shrike and AGM-88 HARM missile strikes against Libyan surface-to-air missile sites at Benghazi.

The squadron then deployed to MCAS Iwakuni, Japan, under the UDP programme, from October 1987 through April 1988, and was assigned to Marine Air Group 15 (MAG-15).

The squadron then returned to the fleet and in August–September 1989 the USS *Coral Sea*, with VFA-132 embarked, operated off the coast of Lebanon. This was in the wake of the Israeli capture of Sheikh Obeid and the killing of Lt Col William R. Higgins, USMC, after he was taken captive in the Lebanon. In early September the squadron provided air cover for the CH-53 Sea Stallion helicopters when evacuating personnel from the US embassy in Beirut. After *Coral Sea* was decommissioned, VFA-132 was reassigned to CVW-6 aboard USS *Forrestal* (CV-59).

In June–September 1991, CVW-6 squadrons participated in Operation Provide Comfort; their role consisted of flying missions over northern Iraq to support the Kurdish relief effort.

The squadron was disestablished on 1 June 1992.

VFA-132	01 Oct 1985–19 May 1986	F/A-18A	CV-43	CVW-13	AK-2xx	Med
VFA-132	Oct 1987–April 1988	F/A-18A	MCAS Iwakuni	AK-2xx	MAG-15	
VFA-132	31 May 1989–30 Sep1989	F/A-18A	CV-43	CVW-13	AK-2xx	Med
VFA-132	Aug 1990–29 Aug 1990	F/A-18A	CVN-72	CVW-13	AK-2xx	WestLant
VFA-132	30 May 1991–21 Dec 1991	F/A-18C	CV-59	CVW-6	AE-3xx	Med

VFA-136 Knighthawks

Akin to a number of other squadrons transitioning to the new Hornet, Squadron One Three Six was established on 1 July 1985 at Naval Air Station Lemoore, California. It received its first F/A-18A on 7 January 1986, before moving rapidly to its new home port of Naval Air Station (NAS) Cecil Field, Florida.

The squadron was first deployed in September 1987 with Carrier Air Wing Thirteen (CVW-13) on board the USS *Coral Sea* (CV43). It joined Carrier Air Wing Seven (CVW-7) on the USS *Dwight D. Eisenhower* (CVN-69) a year later.

Iraq invaded Kuwait on 22 August 1990, during the USS *Dwight D. Eisenhower*'s centennial cruise. The *Eisenhower* was on station in the Red Sea within just thirty-six hours in support of Operation Desert Shield, in the process becoming the first carrier to carry out continued operations in the area. The squadron upgraded to the new Lot 13 Night-attack F/A-18C in November 1990, after it had returned from its deployment, and it became the first fully operational night strike Hornet squadron in the Navy.

Further cruises In October 1991 saw the squadron and the *Eisenhower* returning to the Persian Gulf to assist in enforcing the peace accords set after Operation Desert Storm. The next deployment was aboard the Navy's newest aircraft carrier, USS *George Washington* (CVN-73), for her maiden cruise in May 1994.

In February 1998, VFA-136 embarked on the maiden deployment of USS *John C. Stennis* (CVN-74), a world cruise that Included a tour in the Persian Gulf supporting Operation Southern Watch and concluding with the arrival of the *Stennis* in San Diego, California, its new home port.

The squadron moved to Naval Air Station Oceana immediately afterwards in December 1998, as directed by BRAC decision.

Embarking aboard USS *Dwight D. Eisenhower* to return to the Mediterranean in February 2000, the squadron flew in support of the Dayton Accords, attempting to manage the peace between previously warring parties in Bosnia and around the

Seen during the maiden cruise work up of the USS *George Washington*, F/A-18C 164208 is caught high off the Eastern Seaboard of the United States. It was lost on 12 September 2014 whilst operating with VFA-94.

Balkans. VFA-136 undertook its next cruise in February 2002 aboard the USS *John F. Kennedy* (CV-67) as part of the Sixth Fleet. Two years later it was back aboard CVN-73 once again as part of SACLANT's (Supreme Allied Commander Atlantic) Sixth Fleet. Upon return to NAS Oceana in July 2004 the squadron lost its night-attack role when it was transferred to Carrier Air Wing One (CVW-1) prior to the transition to the new F/A-18E Super Hornet and relocation to NAS Lemoore.

VFA-136 29 Sep 1987–28 Mar 1988	F/A-18A	CV-43	CVW-13	AK-3xx	Med
VFA-136 17 Apr 1989–20 May 1989	F/A-18A	CVN-69	CVW-7	AG-3xx	WestLant
VFA-136 06 Oct 1989–02 Nov 1989	F/A-18A	CVN-69	CVW-7	AG-3xx	WestLant
VFA-136 25 Jan 1990–10 Feb 1990	F/A-18A	CVN-69	CVW-7	AG-4xx	WestLant
VFA-136 08 Mar 1990–12 Sep 1990	F/A-18A	CVN-69	CVW-7	AG-3xx	Med, Arabian Gulf
VFA-136 26 Sep 1991–02 Apr 1992	F/A-18C(N)	CVN-69	CVW-7	AG-3xx	Med
VFA-136 20 May 1994–18 Nov 1994	F/A-18C(N)	CVN-73	CVW-7	AG-3xx	Med
VFA-136 26 Jan 1996–23 Jul 1996	F/A-18C(N)	CVN-73	CVW-7	AG-3xx	Med
VFA-136 26 Feb 1998–26 Aug 1998	F/A-18C(N)	CVN-74	CVW-7	AG-3xx	World Cruise, Arabian Gulf
VFA-136 18 Feb 2000–18 Aug 2000	F/A-18C(N)	CVN-69	CVW-7	AG-3xx	Med, Arabian Gulf
VFA-136 07 Feb 2002–17 Aug 2002	F/A-18C(N)	CV-67	CVW-7	AG-3xx	Med, north Arabian Sea
VFA-136 20 Jan 2004–26 Jul 2004	F/A-18C(N)	CVN-73	CVW-7	AG-3xx	Med, Mid-East
VFA-136 02 May 2006–18 Nov 2006	F/A-18C	CVN-65	CVW-1	AB-3xx	Med, Mid-East, WestPac
VFA-136 07 Jul 2007–19 Dec 2007	F/A-18C	CVN-65	CVW-1	AB-3xx	Med, Mid-East

VFA-137 Kestrels

Strike Fighter Squadron 137 (VFA-137) was established on 2 July 1985, receiving its first Lot 8 F/A-18A Hornet four months later on 25 November 1985. The squadron was nicknamed the Kestrels after the native North American falcon.

In 1987 the squadron made its first deployment to the Mediterranean Sea embarked on the 'Ageless Warrior' USS *Coral Sea* (CV43), as part of Carrier Air Wing Thirteen (CVW-13). The next cruise on board *Coral Sea* saw the squadron patrolling the Eastern Mediterranean after the murder of Lt Col Higgins in Lebanon.

October 1990 saw the squadron transferred to Carrier Air Wing Six (CVW-6) on board USS *Forrestal* (CV-59), and after completing an accelerated preparation cycle it was deployed flying sorties over Iraq in support of Operation Provide Comfort.

A home port change to NAS Lemoore took place in September 1992, which also saw transition to the night-attack capable Lot 15 F/A-18C. In May 1993, the squadron joined Carrier Air Wing Two (CVW-2) and embarked on board USS *Constellation* (CV-64) for its first Western Pacific deployment in November 1994.

During this, and the 1997 deployments, the Kestrels were responsible for patrolling the skies over Iraq, enforcing the UN no-fly zone in support of Operation Southern Watch. The Kestrels used precision-guided ordnance against Iraq in both 1999 and 2001, as part of a Coalition forces response to recurring no-fly zone violations.

November 2002 saw the squadron deployed to the Arabian Gulf on board *Constellation*. This was to be for its last deployment, taking part in widespread operations over Iraq, initially in support of Operation Southern Watch, and following that in combat operations during Operation Iraqi Freedom. During this period of conflict, the Kestrels flew more than 500 combat sorties and dropped more than 300,000lb of precision-guided ordnance.

Block 39 F/A-18C(N) 164715 'NE-406' of VFA-137 seen between WestPac cruises in May 1996. (Photo Robbie Shaw)

Upon returning home in June 2003, transition commenced to the new Lot 25 F/A-18E Super Hornet. After finishing the Safe for Flight certification, the Kestrels found themselves the third F/A-18E squadron in the US Navy.

29 Sep 1987–28 Mar 1988	F/A-18A	CV-43	CVW-13	AK-4xx	Med
31 May 1989–30 Sep 1989	F/A-18A	CV-43	CVW-13	AK-4xx	Med
Aug 1990–29 Aug 1990	F/A-18A	CVN-72	CVW-13	AK-4xx	WestLant
30 May 1991–21 Dec 1991	F/A-18A	CV-59	CVW-6	AE-4xx	Med
27 May 1993–22 Jul 1993	F/A-18C(N)	CV-64	CVW-2	NE-4xx	east coast to west coast
06 May 1994–30 Jun 1994	F/A-18C(N)	CV-64	CVW-2	NE-4xx	RIMPAC
10 Nov 1994–10 May 1995	F/A-18C(N)	CV-64	CVW-2	NE-4xx	WestPac, Persian Gulf
01 Apr 1997–01 Oct 1997	F/A-18C(N)	CV-64	CVW-2	NE-4xx	WestPac, Persian Gulf
18 Jun 1999–17 Dec 1999	F/A-18C(N)	CV-64	CVW-2	NE-4xx	WestPac, Persian Gulf
16 Mar 2001–15 Sep 2001	F/A-18C(N)	CV-64	CVW-2	NE-4xx	WestPac, Persian Gulf
02 Nov 2002–02 Jun 2003	F/A-18C(N)	CV-64	CVW-2	NE-4xx	WestPac, Persian Gulf

VFA-146 Blue Diamonds

On 21 July 1989, VA-146 was re-designated Strike Fighter Squadron One Hundred Forty Six (VFA-146). The squadron received its first F/A-18C(N) Hornet on 18 November 1989, having transitioned from the LTV A-7E Corsair II.

VFA-146 was assigned to Carrier Air Wing Nine (CVW-9), providing the USS Nimitz Air Wing with a night-attack capability, and undertook six cruises aboard the carrier.

In 1997 the squadron undertook a world cruise aboard USS *Nimitz*, and extended its thirteen-year history of more than 55,000 hours without a Class 'A' safety mishap. The climax of the preparation cycle for the cruise was a ninety-six-hour sortie surge operation in which the squadron flew 226 sorties before, on 4 September, departing San Diego with the USS Nimitz Battle Group in support of Operation Southern Watch.

On return to NAS Lemoore, CVW-9 cross-decked to CV-63 for a short cruise to Hawaii before joining CV-62 for the return trip to the United States.

November 2001 saw the squadron deployed with CVW-9 on the USS *John C. Stennis* to conduct combat operations in support of Operation Enduring Freedom over Afghanistan. The 11 September attacks meant the squadron deployed two months early and on 12 December 2001 the squadron began its first night strikes into Afghanistan.

The missions vacillated from four and a half to six hours in conjunction with both the Roosevelt Battle Group and the Kennedy Battle Group. During this time the squadron amassed more than 3,500 flight hours and delivered more than 102,000lb of ordnance. Weapons included laser-guided bombs (LGBs), JDAM and Mark 82 bombs. The squadron returned at the end of May 2002 after a stopover in Australia and Hawaii.

VFA-146 then began a cruise aboard USS *Carl Vinson* (CVN-70) on 17 January 2005. The around-the-world deployment took it across the Pacific and Indian Oceans into the Persian Gulf, where the squadron flew missions in support of Operation Iraqi, along with the rest of CVW-9, for a period of three months. The USS *Carl Vinson* then travelled around the Sinai Peninsula into the Red Sea, through the Suez Canal, the Mediterranean, and finally across the Atlantic.

The carrier air wing was then re-assigned to the USS *John C. Stennis* (CVN-74) in January 2006 and completed two cruises as part of this battle group. The Blue Diamonds then joined CVW-14 aboard the USS *Ronald Reagan* (CVN-76) for the next two cruises, before joining CVW-11 aboard CVN-68.

In 2015, VFA-146 transitioned from the F/A-18C to the newer F/A-18E Super Hornet.

Seen in 2006 when part of CVW-9, F/A-18C 163777 sports a slightly more toned down colourful marking than is normal for a commander's aeroplane.

By 2014, following the last cruise with the legacy Hornet, the commander's aeroplane, in this case 163740, had received a more appropriate scheme. Sadly, the aircraft was destined to be retired shortly afterwards.

One of the last operational flights by a VFA-137 F/A-18C saw this jet, 163496, taken to a place that could be called its true environment, the Sidewinder low-level route. Then passing through Rainbow Canyon, or Star Wars Canyon as it is better known, before dropping out onto Panamint Valley at Point Juliet on the Sidewinder route. It departed for storage at AMARG shortly afterwards.

VFA-146	12 Feb 1990–09 Apr 1990	F/A-18C(N)	CV-64	CVW-9	NG-3xx	San Diego to Norfolk
VFA-146	25 Feb 1991–24 Aug 1991	F/A-18C(N)	CVN-68	CVW-9	NG-3xx	WestPac, Persian Gulf
VFA-146	02 Feb 1993 – 01 Aug 1993	F/A-18C(N)	CVN-68	CVW-9	NG-3xx	WestPac, Persian Gulf
VFA-146	25 Sep 1995–06 Oct 1995	F/A-18C(N)	CVN-68	CVW-9	NG-3xx	FLEETEX, JTFEX
VFA-146	27 Nov 1995–20 May 1996	F/A-18C(N)	CVN-68	CVW-9	NG-3xx	WestPac, Persian Gulf
VFA-146	23 Jun 1997–24 Jul 1997	F/A-18C(N)	CVN-68	CVW-9	NG-3xx	FLEETEX, JTFEX
VFA-146	01 Sep 1997–01 Mar 1998	F/A-18C(N)	CVN-68	CVW-9	NG-3xx	World Cruise
VFA-146	06 Jul 1998–13 Jul 1998	F/A-18C(N)	CV-63	CVW-9	NG-3xx	west coast to Hawaii
VFA-146	Jul 1998–Aug 1998	F/A-18C	CV-62	CVW-9	NG-3xx	Hawaii to west coast
VFA-146	07 Jan 2000–03 Jul 2000	F/A-18C	CVN-74	CVW-9	NG-3xx	WestPac, Persian Gulf
VFA-146	12 Nov 2001–28 May 2001	F/A-18C	CVN-74	CVW-9	NG-3xx	WestPac, north Arabian Sea
VFA-146	13 Jan 2003–19 Sep 2003	F/A-18C	CVN-70	CVW-9	NG-3xx	WestPac
VFA-146	13 Jan 2005–31 Jul 2005	F/A-18C	CVN-70	CVW-9	NG-3xx	WestPac, around-the-world
VFA-146	16 Jan 2007–31 Aug 2007	F/A-18C	CVN-74	CVW-9	NG-3xx	WestPac, Persian Gulf
VFA-146	13 Jan 2009–10 Jul 2009	F/A-18C	CVN-74	CVW-9	NG-3xx	WestPac, north Arabian Sea
VFA-146	02 Jun 2010–08 Aug 2010	F/A-18C	CVN-76	CVW-14	NK-3xx	off the coast of Canada, TSTA, and RIMPAC
VFA-146	02 Feb 2011–09 Sep 2011	F/A-18C	CVN-76	CVW-14	NK-3xx	JTFEX, WestPac, CENTCOM AOR
VFA-146	11 Jun 2012–10 Aug 2012	F/A-18C	CVN-68	CVW-11	NH-3xx	RIMPAC
VFA-146	29 Sep 2012–19 Nov 2012	F/A-18C	CVN-68	CVW-11	NH-3xx	COMPTUEX, JTFEX
VFA-146	19 Apr 2013–12 Dec 2013	F/A-18C	CVN-68	CVW-11	NH-3xx	WestPac, Med, north Arabian Sea

VFA-147 Argonauts

The squadron's last LTV A-7E Corsair II was retired following a Western Pacific, Indian Ocean deployment in February 1989, and Attack Squadron 147 was officially re-designated as Strike Fighter Squadron 147 (VFA-147) on 20 July 1989.

The first Boeing F/A-18C Hornet aircraft arrived on 6 December 1989, with a complete transition to eleven new aircraft seeing the Argonauts become the first squadron to receive the new Lot 12 night-attack Hornets. The first WestPac deployment aboard USS *Nimitz* (CVN-68) to the Persian Gulf took place in March 1991. Combat air patrol missions with CVW-9, the squadron's assigned wing, over Iraq and Kuwait in support of Operation Desert Storm troop withdrawal operations was one of the unit's principle actions during the deployment.

The Argonauts continued leading the F/A-18's maturation as the Navy's premiere strike aircraft by being the first operational night-attack squadron to employ target and navigational forward-looking infrared pods (NAV FLIR) and night vision goggles.

In June 1995 the Argonauts transitioned to new Lot 16/17 versions of the F/A-18C and made three more cruises to the Gulf in support of Operation Southern Watch aboard the USS *Nimitz* (CVN-68).

During May 1998 the squadron traded in its Lot 16 Hornets for an unusual but highly successful transfer to VFA-195 stationed at NAS Atsugi, Yokosuka, Japan. In return, VFA-147 received twelve Lot 11 Hornets, and also a new home aboard the USS *John C. Stennis* (CVN-74). After 11 September 2001, the squadron took part in Operation Noble Eagle, flying combat patrols over Los Angeles.

Operation Enduring Freedom and Anaconda saw the Argonaut pilots flying off the USS *John C. Stennis* (CVN-74) and conducting many strikes into Afghanistan against Taliban and Al-Qaeda forces. During these missions, VFA-147 also helped refine the employment of Joint Direction Attack Munition (JDAM) for the strike fighter community.

With the situation in Iraq escalating in the latter end of 2002, VFA-147, along with the rest of the wing, performed a quick turn around and on 17 January 2003 sailed to the Western Pacific on the USS *Carl Vinson* (CVN-70) for an eight-month deployment. During Operation Iraqi Freedom the squadron delivered more than 120 Mk-80 series weapons, 1,300 Mk-76s, forty LGTRs, six air-to-air missiles, and 12,000 20mm rounds.

In a more standard VFA-147 scheme, 164066 is seen just prior to the squadron's last deployment with the legacy Hornet.

164057, the Argonauts' commander's jet, sporting CVW-9 CAG markings when seen at NAS Lemoore in October 2006.

A further deployment on the USS *Carl Vinson* with CVW-9 in support of Operation Iraqi Freedom took place in January 2005. Two years later the wing cross-decked once again, joining the Stennis Battle Group for an eight-month deployment to the Western Pacific and the Persian Gulf at the beginning of 2007. After the squadron's return it began transition to the F/A-18E.

VFA-147	12 Feb 1990–09 Apr 1990	F/A-18C(N)	CV-64	CVW-9	NG-4xx	San Diego to Norfolk
VFA-147	25 Feb 1991 –24 Aug 1991	F/A-18C(N)	CVN-68	CVW-9	NG-4xx	WestPac, Persian Gulf
VFA-147	02 Feb 1993–01 Aug 1993	F/A-18C(N)	CVN-68	CVW-9	NG-4x	WestPac, Persian Gulf
VFA-147	25 Sep 1995–06 Oct 1995	F/A-18C(N)	CVN-68	CVW-9	NG-4xx	LEETEX, JTFEX
VFA-147	27 Nov 1995–20 May 1996	F/A-18C(N)	CVN-68	CVW-9	NG-4xx	WestPac, Persian Gulf
VFA-147	23 Jun 1997–24 Jul 1997	F/A-18C(N)	CVN-68	CVW-9	NG-4xx	FLEETEX, JTFEX
VFA-147	01 Sep 1997–01 Mar 1998	F/A-18C(N)	CVN-68	CVW-9	NG-4xx	World Cruise
VFA-147	06 Jul 1998–13 Jul 1998	F/A-18C(N)	CV-63	CVW-9	NG-4xx	west coast to Hawaii
VFA-147	Jul 1998–Aug 1998	F/A-18C	CV-62	CVW-9	NG-4xx	Hawaii to west coast
VFA-147	07 Jan 2000–03 Jul 2000	F/A-18C	CVN-74	CVW-9	NG-4xx	WestPac, Persian Gulf
VFA-147	12 Nov 2001–28 May 2001	F/A-18C	CVN-74	CVW-9	NG-4xx	WestPac, north Arabian Sea
VFA-147	13 Jan 2003–19 Sep 2003	F/A-18C	CVN-70	CVW-9	NG-4xx	WestPac
VFA-147	13 Jan 2005–31 Jul 2005	F/A-18C(N)	CVN-70	CVW-9	NG-4xx	WestPac, around-the-world
VFA-147	16 Jan 2007–31 Aug 2007	F/A-18C(N)	CVN-74	CVW-9	NG-4xx	WestPac, Persian Gulf

VFA-151 Vigilantes

On 24 March 1986, the squadron flew off USS *Midway* as VF-151 for the final time, the last flight of the F-4 Phantom II from the deck of an aircraft carrier. The squadron then did the final Phantom TransPac, taking the final four jets direct to Davis-Monthan, with 155505 being the last jet to land. In the meantime, the crews reported to NAS Lemoore, California, for transition to the F/A-18 Hornet. VF-151 was one of just two F-4 fighter squadrons to transition directly to the F/A-18 and was re-designated Strike Fighter Squadron 151 (VFA-151) on 1 June 1986. In November, it returned to the USS *Midway* and NAF Atsugi.

In September 1988 the squadron was embarked on *Midway* operating in the Sea of Japan as a demonstration of US support for a peaceful Olympics as they took place in Seoul, South Korea. A year later, in December 1989, *Midway*, with VFA-151 embarked, was stationed off the coast of the Philippines while that country underwent an attempted coup.

Later in October 1990, in support of Operation Desert Shield, VFA-151 deployed aboard the *Midway* as tensions intensified in Iraq, and on 17 January 1991 the squadron took part in the initial air strikes of Operation Desert Storm, during which it dropped more than 817,000lb of ordnance on key targets in Iraq, Kuwait and the northern Persian Gulf.

The squadron left Japan aboard USS *Midway* bound for NS Pearl Harbor, Hawaii, in August 1991 on the *Midway*'s last under way period. VFA-151 then transferred to CVW-2, and moved to NAS Lemoore, California. In February 1993 the squadron transitioned to the upgraded F/A-18C Hornet Lot 15 aircraft before embarking on USS *Constellation* (CV-64) to bring her from the east coast around Cape Horn, South America, to her new home port of San Diego, California.

The squadron continued to assist with enforcing no-fly zones over southern Iraq during combat zone WestPac deployments in support of Operation Southern Watch in 1994-95, 1997 and 1999 before its final deployment aboard the *Constellation* in November 2002. *Constellation* was then decommissioned in San Diego, California, on 7 August 2003.

VFA-151 Vigilantes was one of only two US Navy former F-4 Phantom operators to transition to the F/A-18A. Always part of CVW-5 and the USS *Midway*, the squadron was re-assigned to CVW-2 upon transition to the night-attack F/A-18C(N). 163777, seen here in the commander's high visibility scheme, was part of the squadron complement until conversion to the Super Hornet in 2013. The aircraft was then flown to Davis-Monthan for storage.

F/A-18C(N) night-attack Hornet 164719 seen at NAS Lemoore in March 2009 between cruises aboard CVN-72.

At that time the Vigilantes, along with CVW-2, moved to the USS *Abraham Lincoln* (CVN-72). In October 2004, the squadron deployed on its first WestPac surge in support of the Navy's new Fleet Response Plan. While deployed, the Vigilantes helped survivors of the Boxing Day tsunami in south-east Asia, participating in Operation Unified Assistance.

Three more cruises took place on the *Lincoln* and in March 2008, the Vigilantes departed on a seven-month Western Pacific Deployment aboard in support of Operations Enduring Freedom and Iraqi Freedom. During this period VFA-151 completed 468 combat sorties and 1,795 deployed sorties, while amassing 2,600 flight hours in direct combat support missions.

The deployment on board USS *Abraham Lincoln* in October 2010 until March 2011 in support of Operations New Dawn and Enduring Freedom saw the squadron completing 398 combat sorties, employing ten GPS precision-guided munitions and approaching 1,467 20mm rounds. In December 2011 the squadron undertook a further five-month deployment for Operation Enduring Freedom in the Fifth Fleet area of responsibility. It returned in August 2012, after twice extending this deployment and employing four precision guided bombs and nine strafing runs totalling 1,718 rounds.

VFA-151 began transitioning to Lot 35/36 F/A-18Es in February 2013, and as this took place the Vigilantes left CVW-2 and moved to CVW-9 attached to the USS *John C. Stennis* on 1 June 2013.

VFA-151	09 Jan 1987–20 Mar 1987	F/A-18A	CV-41	CVW-5	NF-2xx	WestPac
VFA-151	23 Apr 1987–13 Jul 1987	F/A-18A	CV-41	CVW-5	NF-2xx	WestPac
VFA-151	15 Oct 1987–12 Apr 1988	F/A-18A	CV-41	CVW-5	NF-2xx	WestPac, Indian Ocean
VFA-151	18 Oct 1988–09 Nov 1988	F/A-18A	CV-41	CVW-5	NF-2xx	WestPac
VFA-151	21 Jan 1989–24 Feb 1989	F/A-18A	CV-41	CVW-5	NF-2xx	WestPac
VFA-151	27 Feb 1989–09 Apr 1989	F/A-18A	CV-41	CVW-5	NF-2xx	WestPac
VFA-151	31 May 1989–25 Jul 1989	F/A-18A	CV-41	CVW-5	NF-2xx	WestPac
VFA-151	15 Aug 1989–11 Dec 1989	F/A-18A	CV-41	CVW-5	NF-2xx	WestPac, Indian Ocean
VFA-151	20 Feb 1990–06 Apr 1990	F/A-18A	CV-41	CVW-5	NF-2xx	WestPac
VFA-151	02 Oct 1990–17 Apr 1991	F/A-18A	CV-41	CVW-5	NF-2xx	WestPac, Persian Gulf
VFA-151	10 Aug 1991–22 Aug 1991	F/A-18A	CV-41	CVW-5	NF-2xx	Japan to Hawaii
VFA-151	28 Aug 1991–14 Sep 1991	F/A-18A	CV-41	CVW-14	NF-2xx	Hawaii to west coast
VFA-151	27 May 1993–22 Jul 1993	F/A-18C(N)	CV-64	CVW-2	NE-3xx	east coast to west coast
VFA-151	06 May 1994–30 Jun 1994	F/A-18C(N)	CV-64	CVW-2	NE-3xx	RIMPAC
VFA-151	10 Nov 1994–10 May 1995	F/A-18C(N)	CV-64	CVW-2	NE-3xx	WestPac, Persian Gulf
VFA-151	01 Apr 1997–01 Oct 1997	F/A-18C(N)	CV-64	CVW-2	NE-3xx	WestPac, Persian Gulf
VFA-151	18 Jun 1999–17 Dec 1999	F/A-18C(N)	CV-64	CVW-2	NE-3xx	WestPac, Persian Gulf
VFA-151	16 Mar 2001–15 Sep 2001	F/A-18C(N)	CV-64	CVW-2	NE-3xx	WestPac, Persian Gulf
VFA-151	02 Nov 2002–02 Jun 2003	F/A-18C(N)	CV-64	CVW-2	NE-3xx	WestPac, Persian Gulf
VFA-151	15 Oct 2004–04 Mar 2005	F/A-18C(N)	CVN-72	CVW-2	NE-3xx	WestPac
VFA-151	01 Jun 2005–23 Jun 2005	F/A-18C(N)	CVN-72	CVW-2	NE-3xx	NorPac
VFA-151	19 Oct 2005–16 Nov 2005	F/A-18C(N)	CVN-72	CVW-2	NE-3xx	EastPac
VFA-151	27 Feb 2006–08 Aug 2006	F/A-18C(N)	CVN-72	CVW-2	NE-3xx	WestPac
VFA-151	13 Mar 2008–12 Oct 2008	F/A-18C(N)	CVN-72	CVW-2	NE-3xx	WestPac, Persian Gulf
VFA-151	07 Sep 2010–24 Mar 2011	F/A-18C(N)	CVN-72	CVW-2	NE-3xx	WestPac, north Arabian Sea
VFA-151	07 Dec 2011–07 Aug 2012	F/A-18C(N)	CVN-72	CVW-2	NE-3xx	World Cruise

VFA-161 was the other former Phantom operator to transition to the F/A-18A when it left the USS *Midway* in May 1986. After a brief period in inactive status it formed part of the re-established CVW-10 and conducted its work-up aboard the USS *Enterprise* 'Big E'. However, following budget cuts both the air wing and VFA-161 were disestablished in 1987, having spent only a few weeks at sea. This particular jet, 163128, was reassigned to VFA-201 but was lost in an accident on 3 December 1999. (Photo via Geoff Rhodes)

VFA-161 Chargers

VF-161 was one of just two US Navy F-4 Phantom II fighter squadrons to be assigned the VFA designation.

Although it was assigned to CVW-5 aboard the USS *Midway* for much of its career, the squadron relocated from NAF Atsugi to NAS Lemoore in May 1986 for transition training in the F/A-18 Hornet and re-designation as a strike fighter squadron.

Following the transfer of the squadron from CVW-5, and while awaiting transfer to a newly established air wing, the squadron was inactive, stationed at NAS Lemoore. The Chargers were then assigned for a short time to the newly re-established Carrier Air Wing Ten (CVW-10, tail code NM) in June 1987. VFA-161 conducted preparation periods on USS *Enterprise* (CVN-65) and was scheduled to deploy on board USS *Independence*, however following cuts to the budget, CVW-10 was disestablished and VFA-161 was disbanded at the end of 1987.

VFA-161 24 Jul 1987–05 Aug 1987 F/A-18A CVN-65 CVW-10 NM-4xx EastPac

VFA-192 Golden Dragons

VFA-192, following transition to the F/A-18A in 1986, replaced VF-161 aboard the USS *Midway*. The squadron has since undertaken some fifty-two sea-going deployments as part of CVW-5, transitioning to the F/A-18C in 1991 and later to the F/A-18C(N) night-attack version in 1998. 162883 was one of its earliest mounts and was subsequently lost on 18 December 2002 while in use with another unit. (Photo Toshiki Kudu)

Transition of the Golden Dragons to the F/A-18A was completed on 10 January 1986 and the squadron was officially re-designated VFA-192, reflecting the dual-role fighter-attack mission. The Dragons also moved their home port across the Pacific to join the Forward Deployed Naval Forces (FDNF) at NAF Atsugi, Japan as part of Carrier Wing Five (CVW-5).

During the first Hornet cruise for the squadron aboard USS *Midway* (CV-41), it participated in Operation Earnest Will, providing escort air cover for reflagged Kuwaiti oil tankers. Following Iraq's invasion of Kuwait in August 1990, the squadron set sail on board *Midway* for Operation Desert Shield, and opened its account on the night of 17 January 1991 by employing anti-air defence missiles to open Operation Desert Storm.

Throughout the 1990s, VFA-192 would deploy five times in support of Operation Southern Watch, first aboard the USS *Independence* (CV-62), and then aboard the *Kitty Hawk* (CV-63) with upgraded night-attack F/A-18C(N) Hornets.

After the terrorist attacks on 11 September, VFA-192 prepared for an emergency deployment aboard the *Kitty Hawk* to provide close air support for special operations forces, while also providing air defence for the vulnerable US Navy facility on Diego Garcia Island.

In January 2003, VFA-192 was again to lead the main assault of Operation Iraqi Freedom. Over the next nine years the Golden Dragons would deploy several times in support of this operation, becoming one of the most deployed Hornet squadrons within the Navy. Also in that period, the USS *Kitty Hawk* was decommissioned, with the air wing cross-decking initially to the USS *George Washington* (CVN-73), then in 2011 to the USS *John C. Stennis* (CVN-74).

VFA-192	09 Jan 1987–20 Mar 1987	F/A-18A	CV-41	CVW-5	NF-3xx	WestPac
VFA-192	23 Apr 1987–13 Jul 1987	F/A-18A	CV-41	CVW-5	NF-3xx	WestPac
VFA-192	15 Oct 1987–12 Apr 1988	F/A-18A	CV-41	CVW-5	NF-3xx	WestPac, Indian Ocean
VFA-192	18 Oct 1988–09 Nov 1988	F/A-18A	CV-41	CVW-5	NF-3xx	WestPac
VFA-192	21 Jan 1989–24 Feb 1989	F/A-18A	CV-41	CVW-5	NF-3xx	WestPac
VFA-192	27 Feb 1989–09 Apr 1989	F/A-18A	CV-41	CVW-5	NF-3xx	WestPac
VFA-192	31 May 1989–25 Jul 1989	F/A-18A	CV-41	CVW-5	NF-3xx	WestPac
VFA-192	15 Aug 1989–11 Dec 1989	F/A-18A	CV-41	CVW-5	NF-3xx	WestPac, Indian Ocean
VFA-192	20 Feb 1990–06 Apr 1990	F/A-18A	CV-41	CVW-5	NF-3xx	WestPac
VFA-192	02 Oct 1990–17 Apr 1991	F/A-18A	CV-41	CVW-5	NF-3xx	WestPac, Persian Gulf
VFA-192	10 Aug 1991–22 Aug 1991	F/A-18A	CV-41	CVW-5	NF-3xx	Japan to Hawaii
VFA-192	28 Aug 1991–11 Sep 1991	F/A-18C	CV-62	CVW-5	NF-3xx	Hawaii to Japan
VFA-192	15 Oct 1991–24 Nov 1991	F/A-18C	CV-62	CVW-5	NF-3xx	WestPac, Sea of Japan
VFA-192	15 Apr 1992–13 Oct 1992	F/A-18C	CV-62	CVW-5	NF-3xx	WestPac, Persian Gulf

VFA-192	15 Feb 1993–25 Mar 1993	F/A-18C	CV-62	CVW-5	NF-3xx	WestPac, Sea of Japan
VFA-192	11 May 1993–01 Jul 1993	F/A-18C	CV-62	CVW-5	NF-3xx	WestPac
VFA-192	17 Nov 1993–17 Mar 1994	F/A-18C	CV-62	CVW-5	NF-3xx	WestPac, Persian Gulf
VFA-192	19 Jul 1994–29 Aug 1994	F/A-18C	CV-62	CVW-5	NF-3xx	WestPac, Sea of Japan
VFA-192	19 Aug 1995–18 Nov 1995	F/A-18C	CV-62	CVW-5	NF-3xx	WestPac, Persian Gulf
VFA-192	09 Feb 1996–27 Mar 1996	F/A-18C	CV-62	CVW-5	NF-3xx	WestPac
VFA-192	15 Feb 1997–10 Jun 1997	F/A-18C	CV-62	CVW-5	NF-3xx	WestPac, Indian Ocean
VFA-192	23 Jan 1998–05 Jun 1998	F/A-18C	CV-62	CVW-5	NF-3xx	WestPac, Persian Gulf
VFA-192	Jul 1998–17 Jul 1998	F/A-18C	CV-62	CVW-5	NF-3xx	Yokosuka to Hawaii
VFA-192	24 Jul 1998–11 Aug 1998	F/A-18C(N)	CV-63	CVW-5	NF-3xx	Hawaii to Yokosuka
VFA-192	30 Sep 1998–13 Nov 1998	F/A-18C(N)	CV-63	CVW-5	NF-3xx	WestPac, Sea of Japan
VFA-192	02 Mar 1999–25 Aug 1999	F/A-18C(N)	CV-63	CVW-5	NF-3xx	WestPac, Persian Gulf
VFA-192	22 Oct 1999–10 Nov 1999	F/A-18C(N)	CV-63	CVW-5	NF-3xx	WestPac, Sea of Japan
VFA-192	11 Apr 2000–05 Jun 2000	F/A-18C(N)	CV-63	CVW-5	NF-3xx	WestPac
VFA-192	26 Sep 2000–20 Nov 2000	F/A-18C(N)	CV-63	CVW-5	NF-3xx	WestPac, Sea of Japan
VFA-192	02 Mar 2001–11 Jun 2001	F/A-18C(N)	CV-63	CVW-5	NF-3xx	WestPac
VFA-192DET.	01 Oct 2001–23 Dec 2001	F/A-18C(N)	CV-63	CVW-5DET.A	NF-3xx	Operation Enduring Freedom
VFA-192	15 Apr 2002–05 Jun 2002	F/A-18C(N)	CV-63	CVW-5	NF-3xx	WestPac
VFA-192	25 Oct 2002–13 Dec 2002	F/A-18C(N)	CV-63	CVW-5	NF-3xx	WestPac
VFA-192	23 Jan 2003–06 May 2003	F/A-18C(N)	CV-63	CVW-5	NF-3xx	WestPac, Persian Gulf
VFA-192	01 Nov 2003–12 Dec 2003	F/A-18C(N)	CV-63	CVW-5	NF-3xx	WestPac
VFA-192	18 Feb 2004–24 May 2004	F/A-18C(N)	CV-63	CVW-5	NF-3xx	WestPac
VFA-192	19 Jul 2004–07 Sep 2004	F/A-18C(N)	CV-63	CVW-5	NF-3xx	WestPac
VFA-192	10 Feb 2005–28 Mar 2005	F/A-18C(N)	CV-63	CVW-5	NF-3xx	WestPac, Sea of Japan
VFA-192	23 May 2005–20 Aug 2005	F/A-18C(N)	CV-63	CVW-5	NF-3xx	WestPac
VFA-192	24 Oct 2005–12 Dec 2005	F/A-18C(N)	CV-63	CVW-5	NF-3xx	WestPac, Sea of Japan
VFA-192	08 Jun 2006–15 Sep 2006	F/A-18C(N)	CV-63	CVW-5	NF-3xx	WestPac
VFA-192	17 Oct 2006–10 Dec 2006	F/A-18C(N)	CV-63	CVW-5	NF-3xx	WestPac
VFA-192	23 May 2007–21 Sep 2007	F/A-18C(N)	CV-63	CVW-5	NF-3xx	WestPac
VFA-192	21 Oct 2007–27 Nov 2007	F/A-18C(N)	CV-63	CVW-5	NF-3xx	WestPac
VFA-192	18 Mar 2008–04 Apr 2008	F/A-18C(N)	CV-63	CVW-5	NF-3xx	WestPac
VFA-192	15 Apr 2008–12 May 2008	F/A-18C(N)	CV-63	CVW-5	NF-3xx	WestPac
VFA-192	28 May 2008–07 Aug 2008	F/A-18C(N)	CV-63	CVW-5	NF-3xx	Yokosuka to San Diego
VFA-192	21 Aug 2008–25 Sep 2008	F/A-18C(N)	CVN-73	CVW-5	NF-3xx	San Diego to Yokosuka
VFA-192	01 Oct 2008–21 Nov 2008	F/A-18C(N)	CVN-73	CVW-5	NF-3xx	WestPac
VFA-192	20 May 2009–05 Jun 2009	F/A-18C(N)	CVN-73	CVW-5	NF-3xx	WestPac
VFA-192	10 Jun 2009–03 Sep 2009	F/A-18C(N)	CVN-73	CVW-5	NF-3xx	WestPac
VFA-192	06 Jun 2009–23 Nov 2009	F/A-18C(N)	CVN-73	CVW-5	NF-3xx	WestPac
VFA-192	25 Jul 2011–02 Mar 2012	F/A-18C(N)	CVN-74	CVW-9	NG-4xx	WestPac, north Arabian Sea, Persian Gulf
VFA-192	27 Aug 2012–03 May 2013	F/A-18C(N)	CVN-74	CVW-9	NG-4xx	WestPac, North Arabian

Sporting a somewhat original scheme, F/A-18C 163703 of VFA-195 is seen on approach to NAF Atsugi in January 1996. (Photo Toshiki Kudu)

VFA-195 Dambusters

VA-195 was re-designated Strike Fighter Squadron 195 (VFA-195) on 1 April 1985, at which point it began the transitioning process to the F/A-18A. The first of twelve new Lot 8 F/A-18As were delivered in the following October, following which VFA-195 was assigned to the newly established CVW-10.

The Dambusters were to deploy on the USS *Independence* (CV-62), set up to have the old CVW-19 tail code 'NM'. Despite this, CVW-10 was never deployed and VFA-195 became the only squadron of this wing not to be disestablished in 1987, rather VFA-195 was reassigned to CVW-5 aboard the USS *Midway*. When CVW-5's fighter squadrons transitioned from the F-4S Phantom II and the A-7E Corsair IIs to the F/A-18 Hornet, VFA-195 joined the Forward Deployed Naval Forces in Japan. The aircraft began arriving at NAF Atsugi from 14 November 1986.

In 1987, the squadron completed two short Westpac cruises aboard USS *Midway* (CV-41), from January to March and April to July. This involved port calls in Subic Bay, the Philippines, Hong Kong and Sydney. Between November 1987 and February 1988, the squadron took part in Operation Earnest Will, which saw reflagged Kuwaiti tankers taken through the Persian Gulf.

When the Iraqi invasion of Kuwait occurred in August 1990, VFA-195 was deployed aboard the USS *Midway*. During Operations Desert Shield and Desert Storm the squadron flew 564 combat missions, delivering 356 tons of ordnance. It also had the distinction of being the first Hornet squadron to deliver a Walleye II glide bomb in combat.

In August 1991 CVW-5 cross-decked from the USS *Midway* (which was being retired) to the new forward-deployed carrier, the USS *Independence* (CV-62) near Hawaii. At which time the squadron switched to the newer C model Hornet. In March 1996, CVW-5 and VFA-195 went to assist in soothing escalating tension in the Taiwan Strait, a situation that blew up during Taiwan's first direct presidential elections. VA-195 made ten deployments aboard the *Independence* between 1991 and 1998 to the Western Pacific and the Indian Ocean.

From 1998 to 2008 VFA-195 was assigned with CVW-5 to the USS *Kitty Hawk* (CV-63), having taken on the night-attack role with the F/A-18(N) version. The squadron aimed its sights at targets in Afghanistan in 2001 in support of Operation Enduring Freedom, and later in 2003 in the Persian Gulf flew 278 combat sorties in support of Operation Iraqi Freedom, delivering an incredible 179,000lb of precision-guided munitions against military targets in just one month.

After the retirement of the *Kitty Hawk*, CVW-5 and VFA-195 moved to the USS *George Washington* (CVN-73) in August 2008 before the squadron began transitioning to the F/A-18E in late 2010.

VFA-195	09 Jan 1987–20 Mar 1987	F/A-18A	CV-41	CVW-5	NF-1xx	WestPac
VFA-195	23 Apr 1987–13 Jul 1987	F/A-18A	CV-41	CVW-5	NF-1xx	WestPac
VFA-195	15 Oct 1987–12 Apr 1988	F/A-18A	CV-41	CVW-5	NF-1xx	WestPac, Indian Ocean
VFA-195	18 Oct 1988–09 Nov 1988	F/A-18A	CV-41	CVW-5	NF-1xx	WestPac
VFA-195	21 Jan 1989–24 Feb 1989	F/A-18A	CV-41	CVW-5	NF-1xx	WestPac
VFA-195	27 Feb 1989–09 Apr 1989	F/A-18A	CV-41	CVW-5	NF-1xx	WestPac
VFA-195	31 May 1989–25 Jul 1989	F/A-18A	CV-41	CVW-5	NF-1xx	WestPac
VFA-195	15 Aug 1989–11 Dec 1989	F/A-18A	CV-41	CVW-5	NF-1xx	WestPac, Indian Ocean
VFA-195	20 Feb 1990–06 Apr 1990	F/A-18A	CV-41	CVW-5	NF-1xx	WestPac
VFA-195	02 Oct 1990–17 Apr 1991	F/A-18A	CV-41	CVW-5	NF-1xx	WestPac, Persian Gulf
VFA-195	10 Aug 1991–22 Aug 1991	F/A-18A	CV-41	CVW-5	NF-1xx	Japan to Hawaii
VFA-195	28 Aug 1991–11 Sep 1991	F/A-18C	CV-62	CVW-5	NF-4xx	Hawaii to Japan
VFA-195	15 Oct 1991–24 Nov 1991	F/A-18C	CV-62	CVW-5	NF-4xx	WestPac, Sea of Japan
VFA-195	15 Apr 1992–13 Oct 1992	F/A-18C	CV-62	CVW-5	NF-4xx	WestPac, Persian Gulf
VFA-195	15 Feb 1993–25 Mar 1993	F/A-18C	CV-62	CVW-5	NF-4xx	WestPac, Sea of Japan
VFA-195	11 May 1993–01 Jul 1993	F/A-18C	CV-62	CVW-5	NF-4xx	WestPac
VFA-195	17 Nov 1993–17 Mar 1994	F/A-18C	CV-62	CVW-5	NF-4xx	WestPac, Persian Gulf
VFA-195	19 Jul 1994–29 Aug 1994	F/A-18C	CV-62	CVW-5	NF-4xx	WestPac, Sea of Japan
VFA-195	19 Aug 1995–18 Nov 1995	F/A-18C	CV-62	CVW-5	NF-4xx	WestPac, Persian Gulf
VFA-195	09 Feb 1996–27 Mar 1996	F/A-18C	CV-62	CVW-5	NF-4xx	WestPac
VFA-195	15 Feb 1997–10 Jun 1997	F/A-18C	CV-62	CVW-5	NF-4xx	WestPac, Indian Ocean
VFA-195	23 Jan 1998–05 Jun 1998	F/A-18C	CV-62	CVW-5	NF-4xx	WestPac, Persian Gulf
VFA-195	Jul 1998–17 Jul 1998	F/A-18C	CV-62	CVW-5	NF-4xx	Yokosuka to Hawaii
VFA-195	24 Jul 1998–11 Aug 1998	F/A-18C(N)	CV-63	CVW-5	NF-4xx	Hawaii to Yokosuka
VFA-195	30 Sep 1998–13 Nov 1998	F/A-18C(N)	CV-63	CVW-5	NF-4xx	WestPac, Sea of Japan
VFA-195	02 Mar 1999–25 Aug 1999	F/A-18C(N)	CV-63	CVW-5	NF-4xx	WestPac, Persian Gulf
VFA-195	22 Oct 1999–10 Nov 1999	F/A-18C(N)	CV-63	CVW-5	NF-4xx	WestPac, Sea of Japan
VFA-195	11 Apr 2000–05 Jun 2000	F/A-18C(N)	CV-63	CVW-5	NF-4xx	WestPac
VFA-195	26 Sep 2000–20 Nov 2000	F/A-18C(N)	CV-63	CVW-5	NF-4xx	WestPac, Sea of Japan
VFA-195	02 Mar 2001–11 Jun 2001	F/A-18C(N)	CV-63	CVW-5	NF-4xx	WestPac
VFA-195DET.	01Oct 2001–23 Dec 2001	F/A-18C(N)	CV-63	CVW-5DET.A	NF-4xx	Operation Enduring Freedom
VFA-195	15 Apr 2002–05 Jun 2002	F/A-18C(N)	CV-63	CVW-5	NF-4xx	WestPac
VFA-195	25 Oct 2002–13 Dec 2002	F/A-18C(N)	CV-63	CVW-5	NF-4xx	WestPac

VFA-195	23 Jan 2003–06 May 2003	F/A-18C(N)	CV-63	CVW-5	NF-4xx	WestPac, Persian Gulf
VFA-195	01 Nov 2003–12 Dec 2003	F/A-18C(N)	CV-63	CVW-5	NF-4xx	WestPac
VFA-195	18 Feb 2004–24 May 2004	F/A-18C(N)	CV-63	CVW-5	NF-4xx	WestPac
VFA-195	19 Jul 2004–07 Sep 2004	F/A-18C(N)	CV-63	CVW-5	NF-4xx	WestPac
VFA-195	10 Feb 2005–28 Mar 2005	F/A-18C(N)	CV-63	CVW-5	NF-4xx	WestPac, Sea of Japan
VFA-195	23 May 2005–20 Aug 2005	F/A-18C(N)	CV-63	CVW-5	NF-4xx	WestPac
VFA-195	2 4Oct 2005–12 Dec 2005	F/A-18C(N)	CV-63	CVW-5	NF-4xx	WestPac, Sea of Japan
VFA-195	08 Jun 2006–15 Sep 2006	F/A-18C(N)	CV-63	CVW-5	NF-4xx	WestPac
VFA-195	17 Oct 2006–10 Dec 2006	F/A-18C(N)	CV-63	CVW-5	NF-4xx	WestPac
VFA-195	23 May 2007–21 Sep 2007	F/A-18C(N)	CV-63	CVW-5	NF-4xx	WestPac
VFA-195	21 Oct 2007–27 Nov 2007	F/A-18C(N)	CV-63	CVW-5	NF-4xx	WestPac
VFA-195	18 Mar 2008–04 Apr 2008	F/A-18C(N)	CV-63	CVW-5	NF-4xx	WestPac
VFA-195	15 Apr 2008–12 May 2008	F/A-18C(N)	CV-63	CVW-5	NF-4xx	WestPac
VFA-195	28 May 2008–07 Aug 2008	F/A-18C(N)	CV-63	CVW-5	NF-4xx	Yokosuka to San Diego
VFA-195	21 Aug 2008–25 Sep 2008	F/A-18C(N)	CVN-73	CVW-5	NF-4xx	San Diego to Yokosuka
VFA-195	01 Oct 2008–21 Nov 2008	F/A-18C(N)	CVN-73	CVW-5	NF-4xx	WestPac
VFA-195	20 May 2009–05 Jun 2009	F/A-18C(N)	CVN-73	CVW-5	NF-4xx	WestPac
VFA-195	10 Jun 2009–03 Sep 2009	F/A-18C(N)	CVN-73	CVW-5	NF-4xx	WestPac
VFA-195	06 Oct 2009–23 Nov 2009	F/A-18C(N)	CVN-73	CVW-5	NF-4xx	WestPac
VFA-195	18 May 2010–01 Nov 2010	F/A-18C(N)	CVN-73	CVW-5	NF-4xx	WestPac

VFA-201 Hunters

The Hunters transitioned to the F/A-18A in January 1999, trading in its formidable F-14s to supplement the regular forces' requirement for additional Tomcats. As a consequence, the squadron was re-designated VFA-201 to signify its multi-mission capability as a strike fighter squadron. It was the only Navy squadron to transition from the Tomcat to the single-seat legacy Hornet.

VFA-201 was the seventh tactical Navy reserve squadron to be mobilised since the Korean War, but was the only squadron to deploy for combat operations. (During the Vietnam War, three east coast-based and three west coast-based USNR squadrons were activated on 28 January 1968.) After finishing SFARP from its home base in Fort Worth and the Carrier Air Wing Eight co-ordinated strike detachment in Fallon, Nevada, the squadron deployed on board USS *Theodore Roosevelt* (CVN-71) from 6 January to 29 May 2003, and operated over Iraq as part of CVW-8 during the early part of Operation Iraqi Freedom.

The squadron delivered more than 220,000lb of ordnance on Iraq targets, and furthermore, it had an unparalleled 84.6 per cent target acquisition rate. The unit won the Carrier Air Wing Eight landing grade competition for its entire preparation and combat deployment. Nineteen pilots were deployed and of those eighteen were graduates of the Topgun training environment, proving the worth of retaining such experience through the Navy Reserve programme.

VFA-201 was officially deactivated on 30 June 2007. Assets moved to its sister squadron, VFA-204, or other active duty Navy and Marine Corps units.

VFA-201	06 Jan 2003–29 May 2003	F/A-18A+	CVN-71	CVW-8	AJ-2xx	Caribbean, Med

Seen on the hangar lift of the *Theodore Roosevelt* is 162889, once on strength of VFA-151 and more recently VMFAT-101. The jet was retired a year later after returning from this cruise.

A pair of VFA-201 aircraft start up prior to an Operation Iraqi Freedom mission from the deck of the *Theodore Roosevelt* in March 2003 armed with, among other ordnance, the JADAM GPS-guided bomb.

VFA-203's primary role was in the fleet adversary training environment and is only known to have had one at sea deployment as part of CVWR-20 in June 1996. 162892 is seen here at Atlanta sporting a two-tone brown adversary colour scheme in November 2000.

VFA-203 Blue Dolphins

Established in July 1970, VA-203's primary mission at NAS Cecil Field was initially to provide contributory support to the fleet, and to be ready to deploy to an aircraft carrier during a crisis as an activated reserve light attack squadron.

With the demise of the LTV A-7E Corsair II and following its transition to the F/A-18, the unit was to be ready to deploy as an activated reserve strike fighter squadron. A further mission was to act as adversaries to active duty Fleet fighter and strike fighter squadrons being trained for deployment. The squadron was therefore re-designated VFA-203 on 1 October 1989.

In 1993, the squadron began training its pilots in the fleet adversary role. It moved to NAS Atlanta, Georgia, in October 1996 in advance of the 1999 BRAC closure of NAS Cecil Field. The unit was disestablished on 30 June 2004.

F/A-18A 163117 AF-300 of VFA-203 seen at NAS Atlanta in November 2000. This particular aircraft has since been retired and forms part of the Navy heritage collection at Pensacola, Florida.

VFA-203	Jun 1996–Jun1996	F/A-18A	CVN-74	CVWR-20	AF-3xx	WestLant

VFA-204 River Rattlers

The final year of LTV A-7E Corsair II operations in Carrier Air Wing Reserve Twenty (CVWR-20), 1990, was one of the squadron's best years, according to the squadron history. Though greatly impaired by the many aircraft transfers and acceptance inspections, VA-204 flew in excess of its original flight hour programme and undertook seven deployments away from its NAS New Orleans base.

On 28 January 1991, VA-204 hit its ten-year milestone for FOD-free operations, and had previously earned the CVWR-20 nomination for the Battle 'E' (CNO Safety Award). It also received the Chief Naval Officer Safety Award for 1990.

Attack Squadron 204 was re-designated Strike Fighter Squadron 204 on 1 April 1991 and took on its first F/A-18A Hornet that month. This ended twelve years of flying the Corsair II. In March 1993 the squadron officially completed its transition to the F/A-18, and quickly began initial adversary training at NAS Oceana, Virginia. In October it became the

Seen at NAS Atlanta is F/A-18A 162438 in April 1994.

first Reserve Strike Fighter Squadron to deliver adversary support to the active duty fleet during a one-week detachment to NAS Key West.

Traditionally, CVWR-20 has deployed every year to maintain its carrier qualification and fleet readiness, thus enabling individual units to supplement the fleet on an as necessary basis. More latterly, however, it has deployed as an augmented unit within a current air wing since the demise of the traditional training carrier.

VFA-204	Jun 1996– Jun 1996	F/A-18A	CVN-74	CVWR-20	AF-4xx	WestLant
VFA-204	21 Sep 2001–13 Nov 2001	F/A-18A	CVN-68	CVWR-20	AF-4xx	Norfolk to San Diego

VFA-303 Golden Hawks

On 1 January 1984 the squadron was re-designated VFA-303 and relocated to NAS Lemoore, becoming the first reserve squadron to transition to the F/A-18A in the following October. Between September and November 1990, a detachment of the squadron's F/A-18s and personnel, along with its sister unit, VFA-305, joined CVW-11 aboard USS *Abraham Lincoln* (CVN-72) for her transit from Norfolk to Alameda, via Cape Horn. This enabled the units to re-qualify in all aspects of operations at sea

Upon return in November 1990, a detachment of squadron aircraft and personnel deployed to NAWS China Lake in direct support of Operation Desert Shield. Their role was to provide critical real world electronic warfare test and evaluation missions and this required their aircraft to be fully set up to employ electronic warfare, AGM-88 HARM missile and electronic countermeasure suites. In early 1993 the squadron also took on the roles of adversary and fleet support.

It was, however, disestablished on 31 December 1994 after budget cuts.

VFA-303	Jan 1986–Feb 1986	F/A-18A	CV-61	CVWR-30	ND-3xx	off the California coast
VFA-303	Aug 1988–Aug 1988	F/A-18A	CVN-65	CVWR-30	ND-3xx	EastPac
VFA-303DET.2	5 Sep 1990–20 Nov 1990	F/A-18A	CVN-72	CVW-11	ND-3xx	WestLant, EastPac
VFA-303	14 Aug 1992–24 Aug 1992	F/A-18A	CVN-68	CVWR-30	ND-3xx	EastPac

VFA-303 was the first Navy reserve squadron to transition to the F/A-18A, a task it completed in 1985. The squadron, located at NAS Alameda, was however to disestablish in December 1994. This jet, 161708, is now preserved at NAS Fallon, Nevada. (Photo Frank Mirande)

Tasked with the adversary and fleet support role, VFA-305 was the first reserve squadron to undertake total integration with an active carrier air wing when it deployed as part of CVW-11 aboard the USS *Abraham Lincoln* in 1990. Many of the unit's aircraft were to sport the two-tone brown adversary colour scheme.

VFA-305 Lobos

VFA-305 was nicknamed the Lobos from 1974 to 1994. It was a strike fighter squadron component of Carrier Air Wing Reserve Thirty (CVWR-30), and had previously been known as the Hackers. It was first established as Attack Squadron 305 (VA-305) on 1 July 1970 at NAS Los Alamitos, California, part of an overall re-organisation with the mission of increasing the combat readiness of the Naval Air Reserve Force. Later, in January 1971, it moved to NAS Point Mugu.

The squadron was re-designated VFA-305 on 1 January 1987. The squadron was deployed with CVW-11 aboard USS *Abraham Lincoln* from September to November 1990. This was the first total integration of a reserve squadron with an active duty air wing for a lengthy deployment supporting the needs of direct fleet operations.

During this time the Lobos also flew drug interdiction missions and later further added the roles of adversary and fleet support. However, this was short-lived and the squadron was disestablished on 31 December 1994 following a round of budgetary cuts.

VFA-305	Aug 1988–Aug 1988	F/A-18A	CVN-65	CVWR-30	ND-5xx	EastPac
VFA-305	25 Sep 1990–20 Nov 1990	F/A-18A	CVN-72	CVW-11	ND-4xx	WestLant, EastPac
VFA-305	14 Aug 1992–24 Aug 1992	F/A-18A	CVN-68	CVWR-30	ND-5xx	EastPac

VFC-12 Flying Omars

Located at NAF Detroit, Michigan, the squadron was commissioned as VC-12, the Navy's first Reserve Fleet Composite Squadron, on 1 September 1973. The squadron flew the single-seat Douglas A-4 Skyhawk and dual-seat TA-4Js in providing a variety of support services, which included air intercept and dissimilar air combat manoeuvring (ACM) training for Atlantic and Pacific fleet units.

In 1975, VC-12 moved to its present-day home at NAS Oceana, Virginia Beach, Virginia, and in June 1988 was re-designated Fighter Squadron Composite Twelve (VFC-12) to better explain the squadron's mission of dissimilar air combat training (DACT).

In 1994 the unit transitioned to the F/A-18A/B. Ten years later, in 2004, the As were upgraded to the A+. This change meant improved multi-mode mission computers and weapons capabilities. In 2006, VFC-12 traded these F/A-18A+ models (which had low arrested landing fatigue life) for F/A-18Cs from VFA-87. Six years later, in 2012, VFC-12 traded jets again with VFA-87 so it is now once again flying the F/A-18A+.

The squadron's primary mission is to provide support to SFARP, which trains operational fleet F/A-18 squadrons. SFARP is a concentrated three-week training exercise, conducted by the Strike Fighter Weapons School Atlantic, which allows fleet strike fighter aircrews to improve their war fighting skills against a creditable opponent prior to deploying. In addition to the SFARP programme, VFC-12 supports the F/A-18 Fleet Replacement Squadrons at NAS Oceana and with detachments to NAS Key West as well as to NAS Fallon in Nevada, where its distinctively painted aircraft play out the scenario of adversary. It is typical for VFC-12 to spend more than 200 days a year on detachments.

VFC-12, based at NAS Oceana, Virginia, is charged with fleet aggressor training. The unit currently operates F/A-18A+ jets, although has in the past also had F/A-18C mounts on strength. Employing adversary tactics, the squadron has seen a number of 'enemy' schemes over the years. In 2012, when in a period of operating the C model, the choice of scheme was a disruptive blue, as seen on 164258 in this shot.

Transition back to the A+ following a trade with VFA-87 saw a move away from the blue towards a grey scheme in keeping with similar schemes being applied elsewhere. 163113 seen here in Rainbow Canyon, California, during March 2017, still retained such a scheme.

In these shots 162844, 163105 and 163148 can be seen in the current splinter grey scheme that the bulk of the squadron's aircraft have adopted. The nature of their business sees them on a travelling road show working with fleet units on both side of the country. During March 2017 they were detached to NAS Fallon and took time out to hone their low-level training skills.

VFC-13 Saints. The unit only operated the Hornet for a year before relocating to NAS Fallon and transitioning to the Northrop F-5E Tiger II.

VFC-13 Saints

Fleet Composite Squadron Thirteen (VC-13) was established on 1 September 1973 at NAS New Orleans, Louisiana, when the US Navy reorganised the US Naval Reserve and the Naval Air Reserve Force (NAVAIRESFOR). The squadron first flew the Vought F-8 Crusader. Its ranks comprised many former members of VSF-76 and VSF-86 and included seventeen officers and 127 enlisted men. The squadron later made the transition to the Douglas A-4 Skyhawk.

A need for west coast adversary squadrons and other fleet support missions meant the squadron relocated to NAS Miramar, California, in February 1976, and in mid 1976, VC-13 added the two seat TA-4J to the single-seat A-4L in its aircraft complement.

On 22 April 1988, the squadron was re-designated Fighter Composite Squadron Thirteen (VFC-13) and transitioned for a short time to the F/A-18A/B during 1992–93. This improved the squadron's already impressive abilities in ensuring it was presented with a yet more capable and lifelike threat aircraft. When NAS Miramar was transferred back to the US Marine Corps and re-designated as MCAS Miramar after BRAC action, the Navy moved the Naval Fighter Weapons School, or Topgun, to NAS Fallon, Nevada, as part of the Naval Strike and Air Warfare Centre (NSAWC).

VFC-13 transferred to NAS Fallon in April 1996 and soon afterwards made the transition to F-5E Tiger IIs, which it still operates.

2

NAVAL STRIKE AND AIR WARFARE CENTRE (NSAWC)

The office of the Navy describes the Naval Strike and Air Warfare Centre (NSAWC) at Naval Air Station Fallon as the centre of excellence for naval aviation training and tactics development. NSAWC provides service to aircrews, squadrons and air wings throughout the United States Navy through flight training, academic instructional classes, and direct operational and intelligence support. NSAWC flies and maintains a mixture of F/A-18C/D Hornets, F/A-18E/F Super Hornets, E/A-18G Growlers, F-16 Fighting Falcons, Grumman E-2D Hawkeyes and SH-60R/S Seahawk helicopters.

NSAWC is a consolidation of three commands into a single command structure that took place on 11 July 1996, with the aim of enhancing aviation training effectiveness. The Naval Strike Warfare Centre (STRIKE 'U') based at NAS Fallon since 1984, was joined with the Navy Fighter Weapons School (Topgun) and the Carrier Airborne Early Warning Weapons School (Topdome), which both moved from NAS Miramar as a result of a BRAC decision in 1993. NAS Miramar was transferred to the United States Marine Corps to become the west coast home to all USMC F/A-18 Hornet units.

Originally the Naval Strike Warfare Centre, the unit saw three commands consolidated into one in 1996. Prior to this the jets based at NAS Fallon carried the 'Strike', marking as seen on this F/A-18A, 162844, in 1995. (Photo Don Jay)

NAS Fallon also runs the Growler course as well as hosting air wing pre-deployment training.

Currently the unit has only a few early legacy Hornets on strength, such as F/A-18C(N) 16[illegible] seen taxiing in following a Topgun training sortie in 2011.

3

BLUE ANGELS

The Blue Angels is the United States Navy's flight demonstration squadron. It boasts aviators from both the Navy and from the Marines. The team is not only a symbol for the services it represents but also an ambassador for the aircraft manufacturers, whether it was Grumman for the F-11A Tiger, Douglas for the nimble A-4 Skyhawk 'Scooter', or McDonnell for the giant and impressive F-4 Phantom II or today's Boeing F/A-18 Hornet.

The team was formed in 1946, and with its long history it is the second oldest formal flying aerobatic team (under the same name) in the world, behind the French Patrouille de France, which formed in 1931. Some forty years later, on 8 November 1986, the Blue Angels unveiled its present aircraft, the F/A-18, at the same time marking the beginning of its fortieth anniversary year of display flying.

The team has a number of aircraft on strength to allow it to always operate as a 'six pack'. Generally the road show will involve six C model and a pair of two-seat D model aircraft. The team numbers are interchangeable, allowing for aircraft swap should any unserviceability take place. The team is also made up of a mix of personnel drawn from both the US Navy and Marine Corps.

The Blues' Hornets are all drawn from former fleet aircraft that are considered close to being combat-ready, although with a number of modifications including removal of the aircraft gun and replacement with the tank that contains smoke-oil used in demonstrations, and outfitting with the control stick spring system for more precise aircraft control input.

The team transitioned from the first-generation F/A-18A/B version of the Hornet to the F/A-18C/D variant as the A model airframes became life expired.

With the increasing age of the legacy jets, in July 2016 Boeing was awarded a $12 million basic ordering contract to begin converting the F/A-18 E/F for the Blue Angels' use, followed by a further $17,002,107 firm-fixed-price delivery agreement in fiscal 2018 for the retrofit documentation and kits to convert nine F/A-18E and two F/A-18F aircraft into a Blue Angel configuration. The latest estimates are that this transition will not be completed before 2023.

Formed in 1946, the Navy's flight demonstration team transitioned to the current front-line fleet Hornet in 1986 during its fortieth year. Initially flying A models, the team then transitioned to the newer Cas; these became more widely available when the fleet moved into the Super Hornet world.

4

UNITED STATES MARINE CORPS SQUADRONS

VMFAT-101 Sharpshooters

Marine Fighter Attack Training Squadron 101 (VMFAT-101), the Sharpshooters, was commissioned in California at Marine Corps Air Station (MCAS) El Toro on 3 January 1969, as part of Marine Combat Crew Readiness Training Group 10 (MCCRTG-10), 3rd Marine Aircraft Wing (3MAW), flying the McDonnell F-4B Phantom II.

The last F-4 replacement aircrew graduated from VMFAT-101 on 20 May 1987. The following July the squadron flew its remaining ten F-4S aircraft to Davis-Monthan Air Force Base, Arizona, for permanent storage. VMFAT-101 flew the Phantom for eighteen years, during which time the Sharpshooters accumulated more than 125,000 flight hours training Marine and Navy aircrews for the fleet.

On 29 September 1987 VMFAT-101 made ready to stand up as the third F/A-18 Fleet Replacement Squadron (FRS) following on from two US Navy squadrons. MCCRTG-10 deactivated on 31 March 1988 and VMFAT-101 joined Marine Aircraft Group Eleven (MAG-11). By October, the Sharpshooters owned twenty-one F/A-18A/B aircraft. They had trained twenty-five qualified instructor pilots and were about to begin training new Hornet pilots. By May 1989 VMFAT-101 had graduated twenty-three new F/A-18 pilots and had amassed more than 11,000 accident-free Hornet flight hours.

On 10 January 1990, VMFAT-101 saw the arrival of its first two-seat F/A-18Ds and began training rear crew for the transition into the Hornet and by June 1990 the squadron had graduated more than 150 Hornet aircrew, accumulated more than 28,000 hours on F/A-18A, B, C and D versions originating from Lot 6 to Lot 12 aircraft, and just over a month later, on 27 August 1990, Lt Col 'Cajun' Tullos flew the squadron's 50,000th mishap-free flight hour.

VMFAT-101 relocated to MCAS Miramar with the BRAC closure of MCAS El Toro, bringing together all west coast Marine Corps Hornet operations under one roof.

VMFAT-101 also undertakes its own aggressor training as part of it task. To accomplish this but having no dissimilar types in which to fight against, a number of aircraft have in the past received different schemes. Seen here at Fresno in 2014 while on a cross-country navigation exercise is F/A-18B 162864, sporting a tactical disruptive grey scheme. The 'SH' tail code is synonymous with the Sharpshooters, although it has been suggested it stands for something else!

F/A-18B 163115 in a one-off attractive all-over white scheme.

One of the few Marine reserve squadrons to remain active following the Corps' restructuring. VMFA-112 has been activated on a number of occasions, becoming the first reserve squadron to deploy on a Western Pacific exercise since the Korean War. Based at NAS Fort Worth since 1996, this example 162430 is seen on approach to the shared facility in 2011.

VMFA-112 Cowboys

VMF-112 became the Marines' largest reserve squadron upon receipt of the all-weather D/E model of the Vought Crusader, and was later re-designated VMF(AW)-112 and flew several more versions of the F-8 until late 1975, when it switched to the McDonnell-Douglas F-4 Phantom II and was re-designated VMFA-112.

On 28 January 1992 VMFA-112 retired the last active F-4S in Navy service (though some F-4 Phantom IIs were still to be found in Navy testing facilities) and began transition to the F/A-18A. The squadron undertook its first operational flight on 8 October 1992 from NAS Dallas then moved to NAS Fort Worth Joint Reserve Base (JRB) in September 1996, from where it still operates today.

In 2002, VMFA-112's aircraft were upgraded to the F/A-18A+ standard with improved avionics. The Cowboys became the first Reserve squadron to deploy on a Western Pacific exercise since the Korean War when the squadron supported Operation Jungle Shield and Exercise Southern Frontier in the summer of 2004 while operating out of Japan, Guam and Australia.

In 2005 VMFA-112 deployed to Ørland air base, Norway, for the multinational exercise Battle Griffin, which had the aim of improving cohesive operations between multinational forces as well as air-to-ground combat skills while operating in extreme conditions, notably in very cold weather.

VMFA-112 has worked with its sister squadron, VMGR-234, in building the conceptual framework for what has been called the 'Hercules/Hornet Expeditionary Package', a programme planned to allow an F/A-18A+ to land on a 'hasty' runway and then refuel and rearm speedily without the need to return to base, as the present operating procedure demands, and enabling a degree of self-sufficiency that complements the Marine corps' ethos.

In late 2009 the squadron deployed to Al Asad Air Base, Al Anbar Province, Iraq, supporting the ground withdrawal from major Iraqi cities, assisting special forces in a number of missions, as well as providing aerial intelligence. As the final squadron to leave Iraq, it effectively ended active involvement in the Iraq War.

Following its last carrier deployment in 2005 the unit once again undertook its UDP role and was called to arms during operation Iraqi Freedom. F/A-18C 164278 is seen here on a regular weapons detachment to NAF El Centro in October 2015.

VMFA-115 Silver Eagles

VMFA-115 had been flying Phantoms for more than twenty years when it commenced the transition to the F/A-18A on 1 January 1985. The squadron stood up with fourteen aircraft on 16 August the same year, and officially became known as the Silver Eagles in 1986.

Having undertaken numerous UDP deployments with its McDonnell F-4N Phantom IIs in July 1987, VMFA-115 returned to the Western Pacific once again to participate in the Unit Deployment Program at MCAS Iwakuni, Japan, and in 1989 returned to the Philippines and provided assistance to government forces during an attempted coup. The squadron flew armed combat air patrol and escort missions as it helped to calm the situation. Between 1991 and 2000, the Silver Eagles conducted a number of six-month deployments in support of 1st MAW as part of the Western Pacific Unit Deployment Program.

As 2001 drew to a close, the squadron found itself designated a carrier squadron once again, coinciding with the first delivery of the F/A-18A+ aircraft modification. After qualifying all personnel for carrier operations, the squadron deployed with CVW-3 aboard USS *Harry S. Truman* (CVN-71) in October 2002. This culminated in March 2003 with action against Iraqi forces as part of Operation Iraqi Freedom, during which the squadron delivered more than 150 tons of ordnance. A second deployment to the region In October 2004, as part of CVW-3 aboard *Truman*, once again saw the squadron involved in Operation Iraqi Freedom II to assist with Iraq's democratic elections, giving close air support and overhead security.

July 2006 saw the Silver Eagles return to MCAS Iwakuni, Japan, for the first time in seven years as part of the Unit Deployment Program, which included a forward deploy to Kadena air base, Okinawa, and Osan air base in South Korea.

In March 2008, the Silver Eagles deployed to Al Asad, Iraq for Operation Iraqi Freedom. The squadron was the first single-seat forward air control (FAC(A)) Hornet squadron in theatre, providing close air support and overhead security across Iraq. VMFA-115 continues to operate the F/A-18A+ aircraft today from MCAS Beaufort.

VMFA-115	05 Dec 2002–23 May 2003	F/A-18A+	CVN-75	CVW-3	AC-2xx	Med
VMFA-115	02 Jun 2004–25 Jul 2004	F/A-18A+	CVN-75	CVW-3	AC-2xx	Ex Summer Pulse 2004, SACLANT
VMFA-115	13 Oct 2004–18 Apr 2005	F/A-18A+	CVN-75	CVW-3	AC-2x	Med, Persian Gulf

VMFA-121(AW) Green Knights

On 8 December 1989 the squadron was re-designated as VMFA(AW)-121, in the process becoming the first Marine Corps F/A-18D night-attack Hornet squadron. A little over a year later, the Green Knights deployed in support of Operation Desert Shield/Storm, during which they flew 557 sorties and 1,655.5 combat hours in support of the First and Second Marine Divisions helping to liberate Kuwait. This was more than any other Navy or Marine Corps tactical jet squadron.

Returning to MCAS El Toro, California, after hostilities ended, the Green Knights went back to unit deployment rotation and following the closure of the base under the BRAC mandated programme they relocated to MCAS Miramar, California, during August 1994. The Green Knights made three deployments to WestPac before returning to combat over Iraq in March 2000. On this occasion the squadron flew 287 sorties in support of Operation Southern Watch while based at Ahmad al-Jaber Air Base, Kuwait.

After the 11 September attacks, VMFA(AW)-121 was put on the alert under a ninety-six-hour 'prepare to deploy' tether. This remained in force until the unit was called into action in April 2002, after which it deployed with six aircraft to Bishkek, Kyrgyzstan, and the remaining six aircraft and personnel followed them out there a month later. The squadron flew more than 900 combat sorties over Afghanistan in support of Operation Enduring Freedom. It went home to MCAS Miramar in October 2002.

The squadron returned to Ahmad al-Jaber Air Base, Kuwait, just three months after it had returned from Kyrgyzstan. It flew more than 750 combat sorties over Iraq in support of Operations Southern Watch and Iraqi Freedom between February and May 2003. It returned to Miramar on 12 May.

In the planned Marine Corps upgrade programme the squadron became the first Marine Corps unit to transition to the futuristic STOVL Lockheed F-35B Lightning II in 2013 and has since deployed to MCAS Iwakuni, Japan.

VMFA(AW)-121 became the first US Marine Corps F/A-18D night-attack Hornet squadron in 1989. It has now become the first USMC squadron to transition to the fifth generation Lockheed F-35B STOVL Lightning II fighter. Here in March 1992 it can be seen with one of its former mounts, 164046.

F/A-18C 164270, the VMFA-122 commander's aircraft, seen on approach to MCAS Yuma during the autumn 2015 WTI. The squadron will be the next to transition to the Lockheed F-35B, having moved forward in the transition programme due to the growing number of problems with the legacy Hornet fleet.

VMFA-122 The Werewolves

On 14 August 1974, VMFA-122 was placed in a cadre status, in expectation of becoming the Marine Corps' first F-14A Tomcat squadron; then, following the decision not to accept the Tomcat into the Marine Corps inventory, VMFA-122 was reactivated at MCAS Beaufort, South Carolina, and re-equipped with the McDonnell F-4J Phantom II. VMFA-122 flew its final F-4 sortie on 25 September 1985, which saw it completing a full twenty years of service as a Phantom squadron. The following year, in January 1986, the Crusaders – as it was then known as – accepted its first F/A-18A and saw the dawning of a new age.

A number of VMFA-122 training deployments took place across Europe and the US in the 198's, '90s, and into the 2000s. The Crusaders increased their fighting capabilities in October 2001 by transitioning to the F/A-18C model. During this twenty-year period the squadron completed fourteen deployments to the Western Pacific as it participated in the Unit Deployment Program (UDP).

In March 2007 the squadron went back to its Second World War moniker, the Werewolves, in preparation for VMFA-122's first combat deployment since Vietnam. On 29 August 2008 the squadron departed American territory and headed east to participate in Operation Iraqi Freedom.

The unit continued to operate the now aging legacy Hornet for the next ten years, however with issues of legacy Hornet sustainability the Marine Corps pulled an AV-8B Harrier II attack squadron from its spot in the F-35B transition schedule and replaced it with an F/A-18 Hornet fighter-attack squadron because of the better health of the Harrier force.

Marine Attack Squadron 311 (VMA-311), based at MCAS Yuma, Arizona, had been next in line to make the transition in 2018, following on from the transitions of Marine Fighter-Attack Squadron 121 (VMFA-121) and VMA-211.

During a forum with reporters at the Pentagon on 8 February 2017, Lt Gen. Jon Davis said VMFA-122 would take VMA-311's transition slot and, upon completion of F-35B transition, would change its home base from Beaufort to MCAS Yuma.

VMFA-134 Smoke

Re-designated on 1 October 1983, Marine Fighter Attack Squadron 134 (VMFA-134) was a reserve F/A-18 Hornet squadron in the United States Marine Corps. Known as Smoke, the unit was based at MCAS Miramar, California. It fell under Marine Aircraft Group 46 (MAG-46) and the 4th Marine Aircraft Wing (4th MAW) jurisdiction. On 1 April 2007 the squadron was disestablished and transitioned to cadre status, after which its aircraft and personnel were reallocated across the rest of the F/A-18 squadrons.

VMFA-142 Flying Gators

The squadron converted from the Douglas A-4M Skyhawk to the Boeing F/A-18A and on 21 December 1990 it was officially re-designated as Marine Fighter Attack Squadron 142 (VMFA-142).

In August 1997, because of the expected BRAC-mandated closure of NAS Cecil Field by the end of 1999, the squadron moved to Naval Air Station Atlanta in Marietta, Georgia. This followed the reallocation of Naval Air Reserve's Strike Fighter Squadron Two Zero Three (VFA-203) to NAS Atlanta, which had also been based at NAS Cecil Field.

VMFA-142 was absorbed into the Department of the Navy TACAIR concept, which integrated both Marine Corps F/A-18 fighter/attack squadrons and Navy F/A-18 strike fighter squadrons into Navy carrier air wings. As a Marine Air Reserve squadron, VMFA-142 integrated into Carrier Air Wing Reserve Twenty (CVWR-20).

In May 2000 five Marine Hornets from VMFA-142 joined the wing for carrier qualification, qualifying all five pilots in the process. Three of the Hornets subsequently went to Naval Station Roosevelt Roads, Puerto Rico, to undergo further training.

VMFA-142 became the first fixed-wing Marine reserve fighter unit activated to combat since the Korean War when it was deployed to Al Asad Air Base, western Iraq, in support of Operation Iraqi Freedom In February 2005. It assisted in providing combat support in Al Anbar province up until September 2005.

VMFA-142 was placed in cadre status in July 2008, awaiting planned transition to the F-35B Lightning II by 2019, in compliance with a BRAC decision.

Another reserve Marine Hornet unit to fall foul of the restructuring programme, VMFA-134 was assigned as part of MAG-46 at USMC Miramar unit disestablished in 2007. Here 164201, the commander's aircraft, is seen taxiing. (Photo Dan Stijovich)

Another unit placed in 'cadre status' is VMFA-142. Reported to be pending conversion to the F-35B by 2019, the unit was disestablished at NAS Atlanta in July 2008 and its F/A-18A aircraft, including 161965, passed on to other units.

Since converting to the F/A-18, VMFA-212 has participated in Operations Desert Storm and Enduring Freedom while also being one of the units on permanent station at MCAS Iwakuni. However, it also became a casualty in the MCAS reorganisation and was disestablished in 2008 pending transition to the F-35 Lightning II. Here 163775 is seen at Iwakuni in 1997.

VMFA-212 Lancers

The Lancers flew their last McDonnell F-4S Phantom II sorties in August 1988 having achieved more than 23,000 accident-free hours amassed before transitioning to the F/A-18C.

The squadron deployed to Bahrain in December 1990 as part of Operation Desert Shield, from where it conducted air interdiction and close air support missions in support of Coalition forces. On 13 August 1996 the Lancers took off for the last time from MCAS Miramar en route to MCAS Iwakuni, Japan, where they became permanently stationed following a successful six-month UDP cycle.

Following their aerial refuelling aircraft from VMGR-152, the Sumos, the Lancers became the second Marine Corps squadron to deploy after the terrorist attacks of 11 September. The squadron left MCAS Iwakuni on 12 September to begin combat air patrols over Guam supporting Operation Noble Eagle.

In spring 2002 the squadron deployed to Kuwait, when single-seat C models and two-seat Ds from VMFA(AW)-332 the Moonlighters were combined into one unit for the first time. The Moon-Lancers went into Iraq as part of Operation Southern Watch and Afghanistan supporting Operation Enduring Freedom, the latter missions being generally nocturnal and lasting typically for ten hours.

Due to a re-organisation within Marine aviation, the squadron was deactivated in 2008 in order to facilitate the Corps' transition to the F-35 and it was replaced at Iwakuni by VMFA-121.

VMFA(AW)-224 Bengals

The USMC has been a great advocate of the fast FAC night-attack mission and invested heavily in the dual-seat F/A-18D Hornet to fulfil this role. The Bengals, after receiving the aircraft in 1993, have participated in nearly every war and skirmish that the US forces have been involved in. Based at Beaufort, South Carolina, the unit is a regular visitor to NAF El Centro, where it undertakes warm weather and weapons training in the winter months. 164886, seen here, sports the unit's special low visibility tiger-striped marking.

On 6 March 1993, the squadron was re-designated VMFA(AW)-224 and moved to MCAS Beaufort, South Carolina, where it received the multi-mission F/A-18D.

A year later it deployed to Aviano, Italy, as part of the United Nations force for Operations Deny Flight and Provide Promise in Bosnia-Herzegovina. The squadron flew 1,150 sorties for 3,485 flight hours, including 1,150 night hours. In September 1995 the Bengals went back to Aviano again, this time as part of NATO Operations Deny Flight, Deliberate Force and Joint Endeavour. They returned for a third time on 16 February 1997, again in support of NATO-led mandates Operation Deliberate Guard and Operation Silver Wake.

In January 1999, the squadron deployed to MCAS Iwakuni, Japan, as part of the rotational UDP cycle. July 2001 brought yet another UDP to Japan, this time with the introduction of the Advanced Tactical Airborne Reconnaissance System (ATARS), marking the return of the Corps' organic tactical reconnaissance to the Pacific theatre after an absence of more than a decade. The squadron was split into two detachments throughout most of this deployment, which flew in support of both the 31st Marine Expeditionary Unit (MEU) and 15th MEU, conducted ATARS reconnaissance missions, and other such operations as considered necessary by 1 MAW and MAG-12. During the course of the UDP, the squadron operated out of Guam, Okinawa, Australia, Truk Island, Papua New Guinea, the Philippines, South Korea and mainland Japan.

The Western Pacific beckoned again in July 2003, for another UDP. The squadron changed locations five times during this deployment, training with international forces that included Royal Australian Air Force (RAAF) Hornets and Japanese Air Self Defense Force (JASDF) Phantom IIs.

Shortly after the squadron returned from the Western Pacific it was placed on a ninety-six-hour Prepare to Deploy Order (PTDO) in support of Operation Iraqi Freedom. The Bengals turned to and focused exclusively on air-to-ground training.

By July 2004, the squadron had modernised some 75 per cent of its aircraft for the Litening AT targeting pod and participated in DAWEX at Camp Lejeune, highlighting both the squadron's FAC(A) and ATARS capabilities. The imagery exploitation process was exercised, allowing II Marine Expeditionary Force (MEF) to use the ATARS imagery for convoy route planning.

VMFA(AW)-224 deployed to Al Asad Airbase, Iraq, on 11 January 2005 to support Operation Iraqi Freedom. The Bengals brought a surfeit of exceptional capabilities to the table that were unique to the F/A-18D, with the two-seat Hornet providing the General Forward Air Controller and Tactical Air Coordinator additional capabilities. Adding the Pioneer and Predator data link capability to the Litening AT pod, combined with the two-seat Hornet, thus provided the CFACCs and II MEFs JTACs (Joint Terminal Attack Controller) with the ability to be confident in their ability to identify the enemy in real time, reducing the time needed to take action against known aggressors.

While in support of Operation Iraqi Freedom, the Bengals took part in a number of operations in support of Regimental Combat Teams 2 and 8, along with other Marines Corps, Army and Air Force units. In this period the squadron employed 65,225lb of ordnance and flew more than 2,500 sorties and 7,000 hours in direct support of Marine, Army and Coalition ground units.

In August 2005 the Bengals returned to their home station of MCAS Beaufort before beginning the continued rotation of deployments to MCAS Iwakuni, Japan, for UDP and involving various exercises including Foal Eagle, Cobra Gold, and Commando Sling.

VMFA(AW)-224 kicked off 2010 with participation in Exercise Red Flag and a new software upgrade to its aircraft before the squadron deployed again to Iwakuni in support of the UDP and another round of Western Pacific exercises, which continue in the rotational pattern today.

VMFA(AW)-225 Vikings

The Vikings of VMFA(AW)-225 were established on 1 July 1991, at MCAS El Toro, California, flying the F/A-18D Hornet and integrated into the Marine Corps' UDP, deploying to MCAS Iwakuni for six-months at a time. The first of these periods commenced in March 1993. Overseas exercises included tours in Korea, Thailand, Okinawa, Australia, Malaysia, Singapore and Alaska.

In March 1995 the squadron was to move to its new home at MCAS Miramar, California, following the closure of MCAS El Toro under the BRAC-mandated programme. Further unit deployments were to occur in September 1995, March 1996, September 1997, and March 2000, with a deployment to Ahmad al-Jaber, Kuwait, for combat operations during Operation Southern Watch in March 2001.

VMFA(AW)-225 returned to UDP in the March 2002 and it followed in the footsteps of Marines during the Second World War when it crossed through the South Pacific Islands of Iwo Jima and Peleliu and placed the first fighters back on Tarawa, Guadalcanal, and

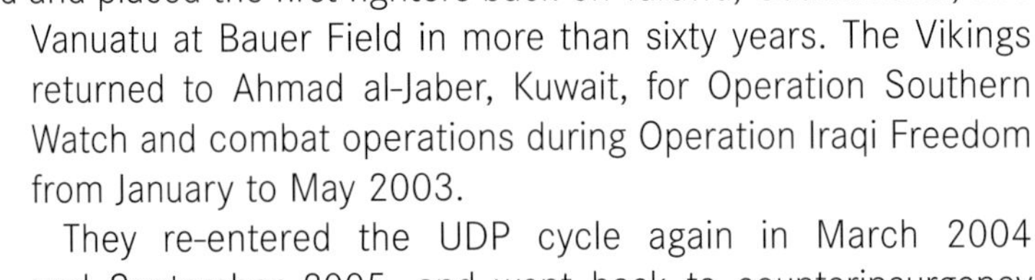

Vanuatu at Bauer Field in more than sixty years. The Vikings returned to Ahmad al-Jaber, Kuwait, for Operation Southern Watch and combat operations during Operation Iraqi Freedom from January to May 2003.

They re-entered the UDP cycle again in March 2004 and September 2005, and went back to counterinsurgency operations (COIN) during Operation Iraqi Freedom, this time operating from Al Asad, Iraq, as the Shepherds from August 2007 to March 2008. The Vikings have subsequently returned to UDP starting in May 2009 and then again in September 2010, subsequently repeating the cycle with other units as and when required.

The west coast sister squadron to the Bengals is VMFA(AW)-225 based at MCAS Miramar. Akin to other non-carrier assigned units, 225 undertakes regular UDP detachments. Seen about to depart its home base in October 2011 is 165417.

VMFA-232 Red Devils

The Marine Corps' oldest active fighter squadron began life on 1 September 1925 at Naval Air Station San Diego, California. By early 1989, the Red Devils had begun the transition to the F/A-18 and by June 1989 the squadron was based at MCAS Kanoehe Bay, Hawaii. December 1990 saw the squadron deploy to Shaikh Isa Air Base, Bahrain, in support of Operation Desert Shield and on 17 January 1991 the Red Devils were among the earliest in crossing the Iraqi border during Operation Desert Storm. During forty-one days of combat operations the squadron completed 740 combat missions totalling 1,390 combat flight hours.

Upon returning to Kaneohe Bay in April 1991, the Red Devils participated in several WestPac deployments and in early 1993 switched their home station to MCAS El Toro, California. They changed it again in 1995, this time back to their birthplace of San Diego, California, at NAS (soon to become MCAS) Miramar.

VMFA-232 was a permanently stationed unit at Kaneohe Bay on Hawaii until 1993, when it returned to mainland United States. Since that time it has become the first Marine Hornet squadron to become part of the Navy's Tactical Aircraft Integration plan. F/A-18C 165186 is seen here exiting the Sidewinder low-level route at Point Juliet over Panamint Valley in October 2014.

The squadron surpassed 50,000 mishap-free flight hours during this time. From 1995 to 2000 the Red Devils continued the excellent precedent they had set by completing eleven Combined Arms Exercises and four UDP deployments while accumulating five Chief of Naval Operations Safety Awards for 1995, 1997, 1998, 1999 and 2000.

After the terrorist attacks of 11 September, VMFA-232 deployed to Andersen AFB, Guam, with the mission of defending both that island and the Marianas Islands. While supporting Operation Noble Eagle, the Red Devils set a new record in Marines fighter aviation when they exceeded 90,000 accident-free flight hours.

As Middle Eastern hostilities escalated, in February 2003 the squadron deployed to Ahmad al-Jaber Air Base, Kuwait, in support of Operation Southern Watch. The Red Devils were among the earliest to fly in the combat zone at the onset of Operation Iraqi Freedom. From March to May 2003, VMFA-232 completed more than 800 combat sorties, totalling 1,700 hours, while dropping more than 640,000lb of ordnance on enemy positions.

Once back at MCAS Miramar in 2004 the Red Devils became the first Marine Hornet squadron to begin the Department of the Navy's Tactical Aircraft Integration Plan. In the latter half of 2004, the squadron completed detachments to NAS Fallon, Nevada, aboard USS *Abraham Lincoln* (CVN-72) with Carrier Air Wing Two, and aboard USS *Nimitz* (CVN-68) with Carrier Air Wing Eleven to take part in its first SFARP, Tailored Ships Training Activity (TSTA), and Composite Training Unit Exercise (COMPTUEX). From May to November 2005 the squadron deployed aboard an aircraft carrier for the first time in forty-six years, operating out of the Arabian Gulf and flying more than 200 combat sorties totalling 1,200 hours in support of Operation Iraqi Freedom.

May 2006 saw the Red Devils transition aircraft once again, this time from the F/A-18C to the F/A-18A+ airframe, which had lower airframe trap time but with upgraded avionics. While the Red Devils now found themselves flying and maintaining the oldest Hornets in the active-duty Navy or Marine Corps, the enormous range of weapons and avionics systems upgrades signified a great improvement in their capability in combat. Soon after this the squadron deployed aboard USS *Nimitz* in April 2007 in support of Operation Enduring Freedom. During the Western Pacific deployment of 2007 the unit amassed 1,124 arrested landings and 2,400 flying hours from the USS *Nimitz* before returning to MCAS Miramar in September 2007.

The squadron was scheduled for another surge deployment to the Western Pacific just four months later as part of the Navy's Fleet Readiness Program. This was once again aboard the *Nimitz*, departing on 2 April 2008, its second Western Pacific cruise in just twelve months.

In February 2010 the Red Devils, having been reunited with the C model Hornet, flew to the Strategic Expeditionary Landing Field in 29 Palms, California, for the Enhanced Mojave Viper exercise. As the first resident F/A-18 squadron in more than five years, VMFA-232 reset the standard for fixed-wing operations. Flying sorties day and night in conjunction with MV-22s, UH-1s and AH-1s, the squadron showcased its talents, not least its great flexibility.

From May to November 2010, the squadron deployed ten F/A-18C and two F/A-18D aircraft to Kandahar Airfield, Afghanistan, in support of Operation Enduring Freedom. As the first land-based F/A-18 squadron in Afghanistan, the Red Devils flew more than 1,700 sorties and 4,900 combat hours. They flew under the call sign Stoic, and in this period the Red Devils dropped more than 71,000lb of ordnance and fired more than 20,000 20mm rounds in support of Marines and Coalition forces in Helmand Province.

In spite of reduced flying hours caused to some degree by sustainability of the legacy Hornet fleet, VMFA-232 still undertook several deployments throughout 2015 and 2016. Sadly, the unit was to lose three aircraft and crews to accidents over a nine-month period, causing a rethink of the F-35 introduction plan.

VMFA-232	07 May 2005–08 Nov 2005	F/A-18C(N)	CVN-68	CVW-11	NH-3xx	WestPac, Persian Gulf
VMFA-232	02 Apr 2007–30 Sep 2007	F/A-18A+	CVN-68	CVW-1	NH-3x	WestPac, Persian Gulf
VMFA-232	24 Jan 2008–03 Jun 2008	F/A-18A+	CVN-68	CVW-1	NH-3xx	WestPac

Another former Kaneohe Bay resident unit was VMFA-235 before returning to initially MCAS El Toro in 1994 then MCAS Miramar in 1996. As part of the Marine Corps realignment programme, the unit was decommissioned later in the year. This jet, 163775, was passed on to VMFA-212 at Beaufort, South Carolina.

VMFA-235 Death Angels

In November 1989, VMFA-235 transitioned to the F/A-18 and after Iraq's 1990 invasion of Kuwait, the squadron went to Bahrain and the Shaikh Isa Air Base to lend its support to Operation Desert Shield. It was the first Fighter squadron to arrive, on 22 August 1990. During the course of Operation Desert Storm it flew more than 2,800 sorties while supporting Coalition forces.

The squadron had moved to MCAS El Toro in 1994 and was attached to Marine Aircraft Group Eleven, but it was reallocated to MCAS Miramar in 1996 following the closure of El Toro under the BRAC mandated realignment plan. However, it was decommissioned later that year on 14 June as part of the Corps' restructuring programme, having only operated the Hornet for a brief seven years.

VMFA(AW)-242 Bats

In August 1990, the Bats, VMF-242, were initially dispatched from MCAS El Toro to the Persian Gulf, but upon reaching Cherry Point, North Carolina, found that the plans had changed. Rather than going to war, they were going back to school for transition to a new aircraft – the F/A-18D night-attack Hornet. They handed off their McDonnell F-4S Phantom II aircraft to other units and returned to El Toro to become Marine All-Weather Fighter Attack Squadron 242, VMFA(AW)-242.

With the new aircraft and new name came a new motto, *Mors ex Tenebris*, Death from the Darkness. The Bats spent the rest of the 1990s training, deploying three times to MCAS Iwakuni, Japan. In 1996, MCAS El Toro closed and the squadron relocated to MCAS Miramar in San Diego, California.

In September 2002, one year after the infamous terrorist attacks but still six months before the US invasion of Iraq, the squadron returned for its eleventh and final deployment to Iwakuni. It remained for thirteen months – the longest such deployment by any squadron since the Vietnam War. Upon their return, the Bats had ten months to prepare for combat. In August 2004, they deployed to Al Asad, Iraq, in support of Operation Iraqi Freedom. In two months, VMFA(AW)-242 delivered more than 300,000lb of ordnance.

The squadron made a second visit to Al Asad from August 2006 to March 2007. One year later VMFA(AW)-242 left on its twelfth deployment to MCAS Iwakuni, but this one was different – it was permanent. After twelve years at Miramar and thirty-eight years in southern California, the Bats moved to Japan and became the Marine Corps' only permanently forward-deployed fighter-attack squadron as the 'resident' squadron of the 'Ready Group'.

F/A-18D 164046 VMFA(AW)-242 'DT-05' seen on approach to NAF Kadena, Okinawa. (Photo Toshiki Kudu)

Originally based out of MCAS El Toro and subsequently MCAS Miramar, VMFA(AW)-242 was to become the only permanently assigned overseas squadron when it relocated to MCAS Iwakuni in 2008, where it remains today. Here one of its night-attack Hornets is seen low in Rainbow Canyon, California. (Photo via Geoff Rhodes)

VMFA-251 Thunderbolts

VMFA-251 flew its last F-4 Phantom in November 1985 after twenty-one years and 85,000 flight hours, and in January 1986 the Thunderbolts transitioned to the Hornet, standing up as the sixth Marine F/A-18A squadron in August the next year. They were able to declare themselves ready for combat in March 1987. Over the following six years, the squadron participated in the Unit Deployment Program (UDP), during which it completed three six-month deployments to Iwakuni, Japan.

From January to April 1994, the squadron supported Operation Deny Flight in Bosnia-Herzegovina, participating in combat missions while deployed to Aviano in Italy. Interestingly, these missions were the first Marine squadron combat sorties in Europe since the First World War.

During May 1994, VMFA-251 received the F/A-18C model Hornet in preparation for becoming a Marine deployable squadron and for assignment to Carrier Air Wing One (CVW 1) aboard the USS *America* (CV 66). The unit successfully completed this first carrier deployment with the F/A-18C in February 1996 supporting Operations Deny Flight and Deliberate Force. The Thunderbolts would deploy twice more with CVW-1, aboard the USS *George Washington* (CVN-73) in September 1997 and the USS *John F. Kennedy* (CV-67) in September 1999. Both deployments were to the Persian Gulf supporting Operation Southern Watch.

Across the summer of 2001 the squadron deployed with CVW-1 aboard the USS *Theodore Roosevelt* (CVN-71), this deployment taking it to the Arabian Sea, where it became the first Marines squadron to have a role in in Operation Enduring Freedom. The squadron went back to MCAS Beaufort in March 2002. A year later, on 12 February 2003, the squadron deployed to Ahmad al-Jaber Airbase, Kuwait, as part of Marine Aircraft Group Eleven, 3rd Marine Aircraft Wing, in support of Operations Southern Watch and Iraqi Freedom.

The squadron flew 560 sorties while it was stationed in the desert, accounting for 1,242.9 flight hours before returning home on 9 May. Further deployments followed as part of UDP. While overseas, VMFA-251 participated in Exercises Cope Tiger, Foal Eagle and Cobra Gold.

On 3 May 2006, VMFA-251 deployed aboard the USS *Enterprise*, signifying the beginning of the 2006 cruise, at which point the ship transited the Atlantic Ocean, Mediterranean Sea and Red Sea on the way to the north Arabian Gulf in support of Operation Iraqi Freedom. Less than five weeks after leaving MCAS Beaufort, the squadron was undertaking combat operations. Its time in the Gulf was short, but the unit still flew 120 sorties, totalling 669 flight hours.

On 24 August 2006, the squadron began a redeployment from the USS *Enterprise* into Al Asad Air Base, Iraq. This was the first occasion in recent years when a squadron had forward deployed from an aircraft carrier to a forward air base. Over the six weeks, the Thunderbolts supported both Operations Iraqi Freedom and Enduring Freedom at the same time. During their sixty-seven days at Al Asad Air Base, the Thunderbolts flew forty-nine consecutive days while supporting Coalition

With the snow-capped Nevada mountains as a backdrop, F/A-18C 164978 is seen on finals to NAS Fallon in November 2014.

VMFA-251 is probably the most deployed-at-sea USMC air squadron. Assigned as part of CVW-1, the unit has deployed at least eleven times with the wing on six different carriers. Sporting a HARM missile and a high-visibility tail flash, 164892 is seen at the 'last chance' at NAS Fallon during the air wing work up in the autumn of 2014.

forces, flying 450 combat sorties, totalling 2,014.7 flight hours in support of Operation Iraqi Freedom. The squadron also flew 187 sorties, totalling 493.5 hours in support of Operation Enduring Freedom. Never before or since has any squadron supported two campaigns simultaneously. The aircraft returned to the ship on 25 October 2006.

On 9 July 2007, the squadron again embarked aboard USS *Enterprise* and began its second combat deployment in just over a year. A month later the air wing would once again provide close air support for Coalition forces in Operation Iraqi Freedom. For three months, the Thunderbolts supported both Operations Iraqi Freedom and Sea Dragon over the shipping routes in the Persian Gulf. The squadron amassed 1,779.5 combat hours in 308 sorties and ensured essential close air support for ground-based Coalition forces.

On 4 November 2007, Carrier Air Wing One (CVW-1) and USS *Enterprise* heard that they were suspending combat operations in Operation Iraqi Freedom and supporting combat operations in Operation Enduring Freedom. This news arrived while they were still flying missions over Iraq. On 6 November 2007, within thirty-six hours of the last CVW-1 aircraft trapping aboard CVN-65, the squadron was flying combat missions in Afghanistan. CVW-1 and the Thunderbolts would support Operation Enduring Freedom for a total of ten days, flying 277.3 combat hours in forty-two sorties. During the cruise, VMFA-251 flew 3,145.6 hours in 1,136 sorties. Some 2,056.8 of those hours and 350 sorties were in combat support.

In May 2009, still as part of CVW-1, the squadron deployed aboard CVN-77, the USS *George H.W. Bush*, for the carrier's shakedown cruise. The air wing including VMFA-251 next deployed aboard the 'Big 'E', CV-65, for two Mediterranean and north Arabian Sea cruises during 2011 and 2012. In March 2015 the squadron joined the USS *Theodore Roosevelt* for a cruise involving the carrier's home port changing from Norfolk to San Diego.

VMFA-251	28 Aug 1995–24 Feb 1996	F/A-18C(N)	CVN-65	CVW-1	AB-2xx	Med, Persian Gulf
VMFA-251	03 Oct 1997–03Apr1998	F/A-18C(N)	CVN-73	CVW-1	AB-2xx	Med, Persian Gulf
VMFA-251	02 Nov 1998–17 Dec 1998	F/A-18C(N)	CVN-75	CVW-1	AB-2xx	Shakedown cruise
VMFA-251	22 Sep 1999–19 Mar 2000	F/A-18C(N)	CV-67	CVW-1	AB-2xx	Med, Persian Gulf
VMFA-251	19 Sep 2001–27 Mar 2002	F/A-18C(N)	CVN-71	CVW-1	AB-2xx	Med, Persian Gulf
VMFA-251	02 May 2006–18 Nov 2006	F/A-18C(N)	CVN-65	CVW-1	AB-2xx	Med, Middle East, WestPac
VMFA-251	07 Jul 2007–19 Dec 2007	F/A-18C(N)	CVN-65	CVW-1	AB-2xx	Med, Persian Gulf, north Arabian Sea
VMFA-251	18 May 2009–30 May 2005	F/A-18C(N)	CVN-7	CVW-1	AB-4xx	Shakedown cruise
VMFA-251	13 Jan 2011–15 Jul 2011	F/A-18C(N)	CVN-65	CVW-1	AB-4xx	Med, north Arabian Sea
VMFA-251	11 Mar 2012–04 Nov 2012	F/A-18C(N)	CVN-65	CVW-1	AB-4xx	Med, north Arabian Sea, Persian Gulf
VMFA-251	11 Mar 2015–23 Nov 2015	F/A-18C(N)	CVN-71	CVW-1	AB-4xx	Norfolk to San Diego

VMFA-312 Checkerboards

VMFA-312 retired its McDonnell F-4 Phantom aircraft in July 1987 before transitioning to the F/A-18A, and by the summer of 1990 it had undertaken its first UDP deployment. This deployment was extended to nine months, from the normal six, as Operations Desert Shield and Desert Storm interrupted the UDP cycle. In April 1991, VMFA-312 returned to Beaufort and in August transitioned to the F/A-18C night-attack Hornet. The summer of 1992 saw VMFA-312 attached to CVW-8 aboard USS *Theodore Roosevelt* (CVN-71). After an extensive nine-month preparation period, the Checkerboards were once again under way in March 1993 for their six-month carrier deployment flying missions over Iraq in support of Operation Southern Watch while operating from the Red Sea.

After returning to Beaufort, VMFA-312 deployed aboard USS *Theodore Roosevelt* (CVN-71) in November 1994 for COMPTUEX and again in January 1995 for FLEETEX training exercises.

As one of the Marine Corps carrier deployable squadrons, it was deployed once again with Carrier Air Wing Eight aboard USS *Theodore Roosevelt* (CVN-71) for its second consecutive Mediterranean deployment in March 1995. During the cruise, VMFA-312 participated in Operations Southern Watch from the Red Sea and Arabian Gulf, then Operations Sharp Guard and Deny Flight from the Adriatic Sea.

The squadron undertook its first direct combat sorties since Vietnam in late August and September 1995, in support of the United Nations resolutions in Operation Deliberate Force. NATO decided to conduct immediate air strikes against Bosnian-Serb ammunition bunkers, communication and control facilities, and logistical storage buildings, which meant intense work for both the air wing and the Checkerboards. The squadron used laser-guided bombs, HARM, and SLAM during both daytime operations and preferred night-time, lights out attacks. The squadron expended three AGM-84s (SLAM), thirteen AGM 88s (HARM), thirteen GBU-10s (2000lb LGB), two GBU-12s (500lb LGB), and fourteen GBU-16s (1,000lb LGB). Some 51,902lb of ordnance was delivered in Operation Deliberate Force. On 21 September 1995 the squadron returned to Beaufort.

The squadron returned to the ship in mid April 1996 for its first deployment in the air wing preparation cycle. Across the rest of 1996, VMFA-312 became familiar with the new air wing, CVW-3. At the end of November 1996, the squadron began another six-month Mediterranean cruise.

A further cruise with Carrier Air Wing Three (CVW-3) aboard USS *Enterprise* (CVN-65) began on 6 November 1998 in support of Operation Southern Watch. The mission would later change as the order came down to commence Operation Desert Fox on 16 December 1998.

The air campaign lasted four nights and would end with the squadron enjoying a 100 per cent sortie completion rate, flying forty-four combat night sorties totalling 120.2 combat hours with 74 per cent of assigned targets destroyed. Some twenty-seven HARM missiles were fired, fifty-three LGBs were dropped and more than 95,500lb of ordnance was loaded.

Following its return from an at sea deployment as part of CVW-3 in 2014, the squadron was reunited with the F/A-18C variant. Here 164975, attending the Topgun course at NAS Fallon, is seen at the runway holding point as Grumman E-2C Hawkeye 168592 of VAW-125 lands.

During this period, VMFA-312 became the first Marine Corps Aviation unit to employ the Joint Stand-off Weapon (JSOW) in combat. With the last Operation Southern Watch mission on 13 April 1999 the Checkerboards had flown 286 combat sorties and 676.7 combat hours in support of Operations Southern Watch, Desert Fox and Deliberate Force.

VMFA-312 began cruise work-ups with Carrier Air Wing Three (CVW-3) in April 2000; the ship in question was the USS *Harry S. Truman* (CVN-75), sailing on her first deployment.

CVW-3 went to the north Arabian Gulf and began Operation Southern Watch missions on 3 January 2001. Less than two weeks later, the Checkerboards gave the USS *Harry S. Truman* an inaugural combat engagement. On 20 January, a VMFA-312 jet destroyed an anti-aircraft artillery site, which had been intimidating Coalition aircraft in southern Iraq. The squadron ended its Operation Southern Watch mission in the April and returned home to MCAS Beaufort on 23 May.

In August 2001, a full pilot swap took place between 312 Checkerboards pilots and 533 Hawk pilots, undertaken by MAG-31. This meant pilots would have to swap over their tactical roles, and in October of that year, MAG-31 would conduct a full aircraft swap, sending the 312 Checkerboards' F/A-18C aircraft to the 122 Crusaders for their F/A-18A aircraft.

May 2002 found the squadron preparing for Transatlantic flight in advance of the multinational NATO exercise Dynamic Mix 02, departing MCAS Beaufort on 18 May and landing eight hours later in Zaragoza, Spain. The squadron deployed six F/A-18As and two F/A-18Ds (augmented by VMFA(AW)-533).

In January 2003, the squadron continued to take the upgraded F/A-18A+ aircraft to the flight line, a change of software that improved its navigational, radar and weaponry suite, enhancing the most technically ferocious of the legacy Hornet fleet. The squadron later deployed aboard the USS *Enterprise* in August 2003. It arrived in the Persian Gulf in October and began support of Operation Enduring and Iraqi Freedom.

In July 2005, the squadron returned to the UDP programme, deploying to MCAS Iwakuni, Japan, before returning to the US for work-ups with Carrier Air Wing Three (CVW-3) during autumn 2006. Upon completion the Checkerboards deployed aboard the USS *George Washington* in late summer 2007.

The next deployment saw the squadron deployed aboard CVN-75 for the final stage of preparation for Joint Task Force Exercise (JTFX), during which it flew joint training missions with both other members of CVW-3 and foreign allies. On return to Beaufort, the Checkerboards performed an impressive air-to-air missile shoot, firing sixteen live missiles and seven training missiles.

The squadron deployed twice more aboard CVN-75 over the next couple of years before transitioning to the F/A-18C. It is currently still assigned under MAG-31 at Beaufort.

VMFA-312	11 Mar 1993–08 Sep 1993	F/A-18C(N)	CVN-71	CVW-8	AJ-330	Med
VMFA-312	22 Mar 1995–22 Sep 1995	F/A-18C(N)	CVN-71	CVW-8	AJ-2xx	
VMFA-312	25 Aug 1996– 22 May 1997	F/A-18C(N)	CVN-71	CVW-3	AC-2xx	Med
VMFA-312	06 Nov 1998–06 May 1999	F/A-18C(N)	CVN-65	CVW-3	AC-2xx	Med, Adriatic, Persian Gulf
VMFA-312	28 Nov 2000–23 May 2001	F/A-18C(N)	CVN-75	CVW-3	AC-2xx	Med, Persian Gulf
VMFA-312	28 Aug 2003–29 Feb 2004	F/A-18A+	CV-63	CVW-1	AB-2xx	Solent, Med, Persian Gulf
VMFA-312	2007–2008	F/A-18A+	CVN-75	CVW-3	AC-2xx	
VMFA-312	21 May 2010–20 Dec 2010	F/A-18A+	CVN-75	CVW-3	AC-2xx	Med, north Arabian Sea
VMFA-312	22 Jul 2013–18 Apr 2014	F/A-18A+	CVN-75	CVW-3	AC-2xx	Med, north Arabian Sea

VMFA-314 Black Knights

In 1982 VMFA-314 received the first of its F/A-18s, and in the process became the first tactical squadron in the Marine Corps and Navy to use the Hornet. VMFA-314 completed carrier qualification later that same year, aboard the USS *Carl Vinson* and the USS *Constellation*.

In 1985 the squadron transferred to Carrier Air Wing Thirteen (CVW-13), and embarked on board the USS *Coral Sea* for duty with the Sixth Fleet in the Mediterranean Sea. The squadron took part in freedom of navigation operations around Libya and combat operations for Operation El Dorado Canyon in the Gulf of Sidra and in Libya.

Two years later the squadron returned to the region when in 1987 it deployed to Egypt as part of joint exercise Bright Star '87 with the Egyptian Air Force. Later, in 1988, the squadron deployed to Balikesir, Turkey, while assigned to MAG-50 in support of the NATO Exercise Display Determination.

UDP followed in 1989 when VMFA-314 returned to the Western Pacific as part of the Unit Deployment Program for the first time since its return from the Republic of Vietnam in 1970. The Black Knights took part in two successive Cope Thunder exercises in the Philippines. They also deployed throughout the Far East in support of Marine Corps front-line units, and went back to MCAS El Toro in October 1989.

In the summer of 1990, while they were getting ready for their next UDP cycle, the Black Knights were deployed swiftly to the Persian Gulf, the first Marine F/A-18 squadron to land in Bahrain for Operation Desert Shield as part of Marine Air Group 70 (MAG-70). The squadron kept up gruelling twenty-four-hour combat air patrols over the Persian Gulf for almost six months.

On 16 January 1991, Operation Desert Shield changed to Desert Storm. The Black Knights flew more than 1,500 hours and 814 combat sorties, which was a greater number than any other Navy or Marine Corps squadron, returning from the Middle East in March 1991.

Shipborne deployment came in August 1992 aboard the USS *Abraham Lincoln* (CVN-72) and participation in Operation Southern Watch, enforcing the UN no-fly-zone in southern Iraq, and in Operation Continue Hope, providing close air support to the 13th and 22nd MEU off the Somalian coast.

In June 1994, VMFA-314 was one of the first MAG-11 squadrons to move from MCAS El Toro to NAS Miramar. In February 1996, the Black Knights took on eighteen new F/A-18C aircraft and, after a successful transition, VMFA-314 transferred to Carrier Air Wing Nine (CVW-9) In 1997 and deployed on an around the world cruise on the USS *Nimitz* (CVN-68). During this time, the squadron took part in Operation Southern Watch again.

January 2000 saw the squadron deployed aboard the USS *John C. Stennis* (CVN-74), with Carrier Air Wing Nine (CVW-9), for a six-month deployment to the Western Pacific and Persian Gulf again in support of Operation Southern Watch. The squadron returned to MCAS Miramar in July 2000 and was reassigned to MAG-11.

The Black Knights are a single-seat Hornet user, that said, at the 2017 El Centro open house the unit turned up with a late production two-seat night-attack F/A-18D, 165532. When questioned the crew confirmed it was 'borrowed' from VMFA(AW)-225 while that unit was undertaking UDP duties. The jet was the last ever legacy Hornet ordered by the Department of the Navy for the USMC, although by default it also received the eight Lot 21 aircraft initially ordered by Thailand.

In an attempt to even out airframes with the most 'traps', VMFA-314 was another unit to transition back from the C model to the upgraded F/A-18A+, this occurring in 2011 and displayed as such at the MCAS Miramar open house the same year.

After the 11 September attacks in 2001, the squadron took part in Operation Noble Eagle in southern California before deploying on 12 November 2001 in support of Operation Enduring Freedom. During this time it delivered more than 69,000lb of ordnance in support of US forces on the ground in Afghanistan. A year later, in January 2003, the Black Knights deployed on board the USS *Carl Vinson* (CVN-70) for a nine-month cruise supporting the Seventh Fleet's mission in the Western Pacific.

In January 2006, the unit retook its place in the Unit Deployment Program and deployed to MCAS Iwakuni, Japan, for six months, where it participated in Exercise Foal Eagle and Northern Edge. This was followed in July 2007 by a further six months to Iwakuni on UDP. The squadron spent four months of this period detached to Kadena AFB in Okinawa.

In March 2009, the Black Knights deployed to Al Asad Air Base, Iraq, supporting Operation Iraqi Freedom. During this time, the squadron flew more than 1,700 combat missions and more than 4,600 flight hours during 179 days of sustained combat operations in Al Anbar Province.

By 2011 the unit had transitioned back to the upgraded F/A-18A+ version and is now scheduled to be the first 'carrier-based' Marine Corps Hornet unit to transition to the Lockheed Martin F-35C Lightning II.

VMFA-314	01 Oct 1985–18 May 1986	F/A-18A	CV-43	CVW-13	AK-3xx	Med
VMFA-314	14 Oct 1992–13 Nov 1992	F/A-18A	CVN-72	CVW-11	NH-2xx	East
VMFA-314	13 Jan 1993–12 Feb 1993	F/A-18A	CVN-72	CVW-11	NH-2xx	East
VMFA-314	02 Mar 1993–28 Mar 1993	F/A-18A	CVN-72	CVW-11	NH-2xx	FLEETEX, East
VMFA-314	15 Jun 1993–15 Dec 1993	F/A-18A	CVN-72	CVN-11	NH-2xx	WestPac, Persian Gulf, Indian Ocean
VMFA-314	23 Jun 1997–24 Jun 1997	F/A-18C	CVN-68	CVW-9	NG-2xx	FLEETEX, JTFEX
VMFA-314	01 Sep 1997–01 Mar 1998	F/A-18C	CVN-68	CVW-9	NG-2xx	World Cruise
VMFA-314	06 Jul 1998–13 Jul 1998	F/A-18C	CV-63	CVW-9	NG-2xx	west coast–Japan transfer
VMFA-314	Jul 1998–Aug 1998	F/A-18C	CV-62	CVW-9	NG-2xx	Japan–west coast transfer
VMFA-314	07 Jan 2000–03 Jul 2000	F/A-18C	CVN-74	CVW-9	NG-2xx	WestPac, Persian Gulf
VMFA-314	12 Nov 2001–28 May 2002	F/A-18C	CVN-74	CVW-9	NG-2xx	WestPac, north Arabian Sea
VMFA-314	13 Jan 2003–19 Sep 2003	F/A-18C	CVN-70	CVW-9	NG-2xx	COMPTUEX, JFTEX, WestPac

VMFA-321 Hell's Angels

In 1991, VMFA-321, which was established on 1 February 1943 at MCAS Cherry Point, North Carolina, stood down the Phantom and began transitioning to the F/A-18 Hornet. The transition was completed six months faster than any active duty squadron, which speaks volumes of the professionalism of the Marines involved.

In April 1994, VMFA-321 took part in the annual Low Country Bombing Derby at the Townsend Target Complex in Georgia. The derby attracted sixty-four tactical aircraft from the active and reserve forces of the Navy, Air Force, and Marine Corps. VMFA-321 took the highest honours in the competition, even though it had to fly the furthest distance to the target area.

Their deployment cycle was rigorous for the active duty Marines, and in an attempt to take some pressure off them, in early 1996 VMFA-321 made the first Marine Reserve transatlantic deployment to Bodø, Norway.

The squadron participated in Exercise Battle Griffin in support of the first exercise accomplished entirely by the Marine Reserves. This was the start of a series of overseas deployments for the Hell's Angels, including three to RAF Leuchars, Scotland, and a joint NATO exercise in support of Bright Star 99/00 at Cairo, Egypt.

The final deployment of the Hell's Angels occurred in June 2004 when the squadron deployed ten F/A-18s to Karup, Denmark, for the NATO Exercise Clean Hunter. This final exercise concluded with concentrated interaction with active duty US forces and Allies, involving deep air strikes and dissimilar air combat with European aircraft.

The squadron was decommissioned on 11 September 2004.

Washington-based VMFA-321 transitioned to the F/A-18A from the venerable F-4N Phantom II in 1991. Aircraft 161979, seen here at Tucson, Arizona, had arrived for some good weather flying following heavy snow on the east coast in January 2000.

VMFA-323 Death Rattlers

F/A-18C(N) 164638 seen landing at MCAS Yuma during the 2014 Weapons Tactics Instructor Course (WTI) exercise.

Marine Fighting Squadron 323 was commissioned on 1 August 1943 at MCAS Cherry Point, North Carolina, gaining its nickname after three fighter pilots killed a 6ft rattlesnake before hanging its skin in the squadron ready room.

On 14 September 1982 the squadron turned in its last F-4 Phantom and officially began the transition to the F/A-18A. In October 1985 it undertook its first seaborne deployment aboard the *Coral Sea* (CV-43), to the Mediterranean on this occasion.

During exercises held in international waters and airspace off the coast of Libya and on 15 April 1986 the squadron was to provide SAM suppression and combat air patrol sorties during the overland strikes on Libyan targets.

VMFA-323 deployed to MCAS Iwakuni, Japan, participating in the six-month USMC Unit Deployment Program for the first time in October 1988. Following a year in MCAS El Toro, California, the Death Rattlers again found themselves in Iwakuni on UDP. As a result of Operations Desert Shield and Desert Storm, this trip was extended to nearly eleven months of being away remaining forward-deployed in Japan to enable other Marine squadrons to deploy to the Persian Gulf.

In April 1993 the squadron took delivery of its first Lot 15 F/A-18C aircraft, the newest model of the now combat-tested Hornet, and they would deploy aboard the USS *Constellation* (CV-64) as a regular component of CVW-2. During two six-month cruises between 1994 and 1997 the squadron flew from *Connie*'s flight deck in support of Operation Southern Watch, enforcing the United Nations sanctions against Iraq and patrolling the no-fly zone south of Baghdad.

Between January 2005 and July 2009 the squadron became a permanent component of Carrier Air Wing Nine (CVW-9) deploying three times, initially aboard the USS *Carl Vinson* (CVN-79) and then the USS *John C. Stennis* (CVN-74). By September 2012 the unit had transferred to CVW-11 for two cruises aboard the USS *Nimitz* (CVN-68) before becoming a component of CVW-14 aboard the USS *Ronald Reagan* (CVN-76).

VMFA-323	01 Oct 1985–19 May 1986	F/A-18A	CV-43	CVW-13	AK-4xx	Med
VMFA-323	06 May 1994–30 Jun 1994	F/A-18C(N)	CV-64	CVW-2	NE-2xx	RIMPAC 94
VMFA-323	10 Nov 1994–10 May 1995	F/A-18C(N)	CV-64	CVW-2	NE-2xx	WestPac, Indian Ocean, Persian Gulf
VMFA-323	01 Apr 1997–01 Oct 1997	F/A-18C(N)	CV-64	CVW-2	NE-2xx	WestPac, Indian Ocean, Persian Gulf
VMFA-323	18 Jun 1999–17 Dec 1999	F/A-18C(N)	CV-64	CVW-2	NE-2xx	WestPac, Persian Gulf
VMFA-323	16 Mar 2001–15 Sep 2001	F/A-18C(N)	CV-64	CVW-2	NE-2xx	WestPac, Persian Gulf
VMFA-323	02 Nov 2002–02 Jun 2003	F/A-18C(N)	CV-64	CVW-2	NE-2xx	WestPac, Persian Gulf
VMFA-323	13 Jan 2005–31 Jul 2005	F/A-18C(N)	CVN-70	CVW-9	NG-2xx	JFTEx, WestPac, World Cruise
VMFA-323	16 Jan 2007–31 Aug 2007	F/A-18C(N)	CVN-74	CVW-9	NG-2xx	WestPac, Persian Gulf
VMFA-323	13 Jan 2009–10 Jul 2009	F/A-18C(N)	CVN-74	CVW-9	NG-2xx	WestPac, west Arabian Sea

VMFA-323	29 Sep 2012–19 Nov 2012	F/A-18C(N)	CVN-68	CVW-11	NH-4xx	COMPTUEX, JTFEX, East Pacific
VMFA-323	30 Mar 2013–16 Dec 2013	F/A-18C(N)	CVN-68	CVW-11	NH-4xx	WestPac, Med, north Arabian Sea
VMFA-323	02 Feb 2011–09 Sep 2011	F/A-18C(N)	CVN-76	CVW-14	NK-4xx	JTFEX, WestPac, CENTROM AOR
VMFA-323	2017–2018	F/A-18C(N)	CVN-68	CVW-11	NH-4xx	

VMFA(AW)-332 Moonlighters

Initially commissioned in June 1943, some fifty years later on 16 June 1993 the Moonlighters were re-designated Marine All Weather Fighter Attack Squadron 332 (VMFA(AW)-332), at the same time relocating to MCAS Beaufort, South Carolina, and transitioning to the F/A-18D.

Barely a year later, the squadron was called on to support NATO's Operation Deny Flight and Operation Provide Promise in the former Yugoslavian republic of Bosnia, deploying to Aviano Air Base, Italy, from October 1994 to March 1995. During the deployment, the squadron led the largest NATO air strike to date against the Udbina Airfield, as well as taking part in a number of other strikes. The squadron deployed a second time to Aviano in March 1996 supporting Operations Decisive Edge and Joint Endeavour.

May 1999 saw the squadron continue its almost nomadic existence when it deployed to Taszár Air Base, Hungary, in support of Operations Allied Force and Joint Guardian. In the period 28 May and 7 June 1999, the Moonlighters flew 120 combat sorties over the Federal Republic of Yugoslavia, performing every conceivable mission for an F/A-18D squadron during both day and night, and in the process expending 175,000lb of ordnance. VMFA(AW)-332 was, along with VMFA(AW)-533, one of the first deployable units flying appropriately configured Hornets for the Joint Direct Attack Munition (JDAM). During the deployment, the squadron exploited the Advanced Tactical Airborne Reconnaissance System (ATARS) operationally for the first time. The squadron returned to MCAS Beaufort in July 1999.

In January 2000 the squadron returned to its UDP responsibilities and deployed to MCAS Iwakuni, Japan, as part of the programme. The deployment would see the squadron visit mainland Japan, Okinawa and Korat, Thailand. On 1 March 2000, the Moonlighters hit 81,000 mishap-free flight hours, a safety record that encompassed twenty-one years.

The next overseas deployment saw the squadron at Andersen Air Force Base, Guam on 11 January 2002 and taking over the Operation Noble Eagle mission. This deployment relieved VMFA-212 and VMFA-232. During the squadron's tour in Guam, it flew 262 sorties and amassed 582.7 hours of flight time.

Later, VMFA(AW)-332 joined VMFA-212 to form two composite squadrons. This gave the Lancers the capability to conduct ATARS reconnaissance missions and FAC(A) in support of Operations Enduring Freedom and Southern Watch. The squadrons exchanged four aircraft and air and ground crew.

The following year the squadron returned to Japan as part of the UDP, during which time it deployed to Dhaka, Bangladesh, supporting Operation Sumo Tiger as well as to Thailand, Guam, and Townsville, Australia.

A former Grumman A-6E Intruder squadron, VMFA(AW)-332 transitioned to the night-attack F/A-18D in 1993. Based out of MCAS Beaufort, the unit undertook a great number of policing duties throughout the world, as well as participating in Operation Iraqi Freedom. (Photo USMC)

A further Operation Iraqi Freedom deployment took place on 26 July 2005. During this time the squadron frequently flew twelve-sortie Air Tasking Order cycles, seeing sporadic surge days as well as ground alerts. Their typical missions included close air support, armed reconnaissance (AR), infrastructure intelligence, surveillance and reconnaissance (ISR), Advanced Tactical Airborne Reconnaissance System (ATARS) and convoy escort.

The squadron flew 2,406 sorties for a total of 6,031.9 hours across seven months of combat operations. This included innumerable exercises, using the call-sign 'dealers'. The squadron left Al Anbar province on 11 February 2006, having expended for some fifty GBU-38s, 270 GBU-12s, twenty-three AGM-65s, ten 5in rockets, two GBU-16s, six GBU-32s, sixteen LUU-19s and 7,640 rounds of 20mm; total of 160,966lb of ordnance.

In late 2006 the squadron began downscaling in readiness for decommissioning and flew its final flight in the F/A-18 in early 2007. The squadron was deactivated on 30 March 2007 holding the lengthiest streak of mishap-free flight hours for any tactical jet squadron at an extraordinary 109,000 hours.

VMFA-333 Fighting Shamrocks

VMFA-333 was the last regular Marine squadron to operate the F-4S Phantom II but eventually transitioned to the F/A-18A in October 1987. Its use of the Hornet was relatively short-lived but during its final deployment to the Persian Gulf in 1990–91 as part of Operation Desert Storm, the Shamrocks flew upwards of 700 combat missions without loss. They also delivered more than 2 million pounds of ordnance against Iraqi forces. VMFA-333 was deactivated effective 31 March 1992 as part of the Corps' continuing structure realignment.

VMFA-451 Warlords

In 1984, VMFA-451 set a safety record of 29,000 mishap-free hours in the F-4 Phantom. It spent twenty-one years in the honoured 'Rhino', under the leadership of Alfred Cunningham award winner Lt Col T.D. Seder. Following this, the Warlords swapped their F-4Ss for the F/A-18A in September 1987. In May 1989, the squadron sailed with the USS *Coral Sea* (CV-43) for a Mediterranean Cruise, ending eight years during which no Beaufort-based Marine units appeared on carriers. During the cruise, VMFA-451 was awarded the 1988 Hanson award as the outstanding Marine fighter-attack squadron, as it exceeded 40,000 mishap-free hours. The squadron returned to Beaufort in September 1989.

On 23 August 1990, VMFA-451 deployed to Bahrain in support of Operation Desert Shield, becoming the first Marine unit to attack Iraqi forces in Operation Desert Storm at 0300 on 17 January 1991. The commanding officer, Lt Col A.S. 'Scotty' Dudley, led a four-plane strike as part of a larger fifty-aircraft strike force under Marine Aircraft Group Eleven. The Warlords ended the operation with 770 combat sorties and 1,400 combat hours and went back to MCAS Beaufort with all aircraft and personnel.

VMFA-451 spent its final six years before deactivation on the east coast. During this period it made intermittent deployments, including exercises in Norway. The squadron returned from the Western Pacific in late July 1996, and so ended its impressive fifty-seven-year story. The Warlords of VMFA-451 were deactivated on 31 January 1997.

VMFA-451	31 May 1989–30 Sep 1989	F/A-18A	CV-43	CVW-13	AK-1xx	Med

VMFA-333 Fighting Shamrocks only operated the F/A-18A for five short years, succumbing to the Marine Corps structure realignment programme in March 1992. Here, 161928 is seen at MCAS Beaufort in 1990. (Photo Robbie Shaw)

F/A-18A 163131 of VMFA-451 seen at its home base of MCAS Beaufort in 1990. The squadron participated in Operation Desert Storm but failed to survive the Marine Corps realignment programme and was deactivated in 1997 after ten years of Hornet operation. (Photo Robbie Shaw)

Seen landing at Kadena AFB, Okinawa, in 1991 is 162414, VMFA-531 was to deactivate a year later as part of the Corps' consolidation of assets.

VMFA-531 Grey Ghosts

On 29 May 1984, VMFA-531 received its first Boeing F/A-18A Hornet, becoming the third fleet squadron to receive the new fighter. Its operation of this versatile aircraft was, however, short-lived and the squadron was decommissioned on 27 March 1992 as part of a consolidation of assets within the Corps structure.

VMFA(AW)-533 Hawks

The Hawks were commissioned as Marine Night Fighter Squadron 533 (VMF(N)-533) at MCAS Cherry Point, North Carolina, in October 1943. By 1 September 1992, 533 had introduced the new F/A-18D, and with it came a move to MCAS Beaufort, South Carolina. This made the unit the first all-weather fighter attack squadron in 2nd Marine Aircraft Wing.

Less than a year later, VMFA(AW)-533 deployed to Aviano, Italy, in July 1993, a location it was to return to three times over the next five years, flying a wide variety of missions to support NATO operations. Following its first UDP in Hornets in 1999, 533 returned one last time to Eastern Europe, this time operating from Taszár, Hungary.

January 2001 to July 2001 saw a second UDP deployment, this time as part of the 31st MEU Special Operations Capable (SOC). The Hawks, as they were known, became the first land-based fixed-wing squadron to successfully execute a MEU (SOC) afloat, involving an island-hopping campaign throughout the Pacific that would take them to airfields not seen by US forces since the Second World War. Instead of undertaking its third scheduled UDP deployment in January 2003, the squadron would instead deploy to the Central Command Area of Responsibility (CENTCOM AOR) supporting Operations Southern Watch and Iraqi Freedom, arriving at Ahmad Al-Jaber Air Base, Kuwait, in 11 February 2003.

F/A-18D 164679 of VMFA(AW)-533 landing at MCAS Yuma during the 2015 autumn WTI exercise. Note the ATARS nose on the jet, which is interchangeable as required.

On 20 March Coalition forces began the ground offensive, supported by the squadron's Hornets. Undertaking twenty-four-hour operations, the squadron expended more than 800,000lb of ordnance, flying 558 sorties totalling 1,440 flight hours.

In February 2006 the squadron deployed to Al Asad Air Base in the Al Anbar Province of Iraq, assigned to the 3rd Marine Aircraft Wing in support of 1st Marine Expeditionary Force. The Hawks employed the F/A-18D with the Litening II FLIR/TV pod in a myriad of roles, which included reconnaissance, surveillance, convoy escort, close air support, strike missions, forward air controller airborne (FAC(A)) and tactical air controller airborne (TAC(A)). Operating around the clock, the squadron again expended more than 110,000lb of ordnance, flying 2,480 sorties and completing 7,456 flight hours.

Today the squadron is still operating the two-seat F/A-18D and is an integral part of MAG-31 at MCAS Beaufort.

5

TEST AND EVALUATION SQUADRONS

VX-4 The Evaluators

VX-4 (Air Test and Evaluation Squadron 4) was established in 1950 at NAS Atlantic City as Air Development Squadron 4, primarily for the development of airborne early warning systems (AWACS). The squadron moved to NAS Patuxent River in 1951; it was disestablished there later the same year as a result of a lack of assigned projects.

VX-4 was re-established in 1952 at NAS (then PMTC) Point Mugu to conduct evaluations of air-launched guided missiles as assigned by the commander, Operational Test and Evaluation Force. In 1960 the squadron began to include extra projects that were not related to guided missiles. ALR-67 (ECP-510) testing in the F/A-18 was one of the squadron's more significant projects in late 1991 and 1992.

By late 1993, Hornet testing was slowing down at VX-4 and preparations had begun for moving F/A-18 projects and F/A-18C/D model aircraft to VX-5 at China Lake as a consolidation with VX-5 began. The first Hornet departed for China Lake before the end of 1993, with the rest of the squadron's aircraft following in early 1994 after AMRAAM testing was completed.

On 30 September 1994 VX-4 was disestablished and its assets were reassigned to VX-9 Vampires, which had been created from the VX-4 and VX-5 amalgamation.

Testing and evaluation of Hornet-related systems was one of the more significant projects being carried out by VX-4 in the early 1990s from its NAS Point Mugu facility. The nearby Pacific Missile Test Centre (PMTC) gave maximum return per sortie but with such projects winding down the decision was taken to relocate the unit's taskings to the NWC at NAS China Lake. F/A-18A 161945 is seen here at Point Mugu early in the Hornet's test phase. (Photo Hans J. Shroder)

VX-5 Vampires

The sister unit to VX-4 at Point Mugu was VX-5, located within the structure of the NWC at China Lake. VX-5 and VX-4 were to amalgamate in 1993. 161523 seen here was retired to AMARG shortly afterwards. (Photo via Andy Thomson)

Air Development Squadron Five (VX-5), was commissioned on 18 June 1951 at NAS Moffett Field, California. In July 1956 VX-5 moved to the Naval Air Facility at China Lake, California, due to its better ranges and instrumentation facilities.

In June 1993, the Chief Naval Officer (CNO) directed the consolidation of VX-4 and VX-5 into a single operational test and evaluation squadron designated VX-9, although retaining a permanent F-14 Tomcat Detachment located at PMTC Point Mugu, California. According to the official Navy history, this initiative was launched as part of the 'right-sizing' of naval forces in the wake of the Cold War.

VX-9 Vampires

In June 1993, the Chief of Naval Operations directed that VX-4 at NAS Point Mugu, California, and VX-5 should to be combined into a single operational test and evaluation squadron to be designated VX-9, at NAWC China Lake. At that time VX-4 was to retain a permanent F-14 Tomcat Detachment located at Point Mugu until such time as that programme came to an end.

The choice of the VX-9 designation is relatively obvious as it combines the 4 and 5 titles of the amalgamating units. The Vampires undertake operational evaluation of systems and weapons, making good use of the vast China Lake range complex. The squadron's role is to prove the systems that have been developed and trialled by its sister squadron, VX-31.

Today VX-9 operates a mixture of legacy and Super Hornets, including the electronic suppression Growler. It also maintains a detachment at Edwards AFB operating the F-35C Lightning II.

The amalgamation of VX-4 and VX-5 into a single unit saw the creation of VX-9, although the squadron retained the VX-5 Vampires title. Today it operates five variants of the Hornet, including a number of legacy series examples. Here F/A-18D 164254 sporting the Vampires special marks, albeit in low-visibility grey, is seen passing through Rainbow Canyon on the 'Jedi Transition' portion of the Sidewinder low-level route.

In 2002 it was decided to restructure the units within the NATC/NAWC umbrella with the Naval Strike Aircraft Test Squadron being re-designated VX-23. Here F/A-18A 161709 is seen in NAWC Salty Dogs colours.

VX-23 Salty Dogs

The squadron was established on 22 July 1995 as the Naval Strike Aircraft Test Squadron at NAS Patuxent River and was re-designated VX-23 on 1 May 2002. Over its history, VX-23 has tested and evaluated fixed-wing fighter, attack and other designated aircraft including EA-6B, F-14, F/A-18 and T-45. In recent times, VX-23 accommodated and provided flight test aircrew for the new Joint Strike Fighter X-32 and X-35 concept demonstration aircraft, as well as the X-31 VECTOR test vehicle, and today supports the Lockheed F-35B and F-35C programmes.

According to the station base guide, the squadron's mission is to support the RDT&E (Research Development Test & Evaluation) of fixed-wing tactical aircraft by providing aircraft and pilot assets, maintenance services, safety oversight and facility support for these efforts. Primary areas of support include flying qualities and performance evaluations, shipboard suitability, propulsion system testing, tactical aircraft mission system testing, ordnance compatibility and ballistic efforts, reliability and maintainability assessments, flight fidelity simulation and flight control software development.

On Earth Day 2010, a modified F/A-18 Super Hornet was demonstrated by the squadron. Called the 'Green Hornet', it used a mix of biofuel made from the camelina sativa plant. More recently it has been involved in the development of the new catapult system being deployed aboard the US Navy's latest generation of aircraft carriers.

VX-31 also operates high-visibility F/A-18C 165210, which is seen here at a MCAS Miramar open house.

The Navy and Marines final test unit in respect of the Hornet is currently VX-31, known as the Dust Devils. Based at NAS China Lake, the unit undertakes concept trials on new equipment and software before passing the task of operational testing on to VX-9. The USMC has begun to consolidate all front-line testing under one umbrella at MCAS Yuma. The MV-22s and UH-1Y/AH-1W/AH-1Z types have already relocated under VMX-1, but whether the Hornets will join them is not clear. Today, however, the unit continues to operate a number of jets within its establishment, including this F/A-18D, 164279.

VX-31 Dust Devils

Air Test and Evaluation Squadron 31 (VX-31 or Airtevron Three One) is a United States Navy air test and evaluation squadron. Based at Naval Air Weapons Station China Lake, California, it uses the tail code DD and flies a variety of fixed-wing aircraft and helicopters for the Navy and Marine Corps. Formed on 24 April 2002 to deliver 'warfighter' capability at the best value, the unit undertakes trials of new systems and sub-systems, including software development, before passing the task of operational testing over to its sister unit, VX-9.

6

NASA

Three F/A-18 Hornet aircraft are reported to be currently flown at NASA's Armstrong Flight Research Centre at Edwards, California, for purposes of research support and pilot proficiency. The current fleet of aircraft was obtained from the US Navy between 1984 and 1991; one being a two-seat B model and others single-seat As. These aircraft are usually called chase planes and take on the role of escort aircraft during research missions, although they often undertake more specific test duties.

NASA has, however, over a period of time operated or used at least fifteen Hornet airframes. Those involved have generally operated with a NASA 'civilian' registration, as detailed below. Some, however, may have only been used for a short period of loan and as such retained their military guise.

When used as systems chase aircraft, the chase pilots are in continuous radio contact with research pilots and help maintain total flight safety during specific tests and manoeuvres.

All chase aircraft, the Hornets included, are employed as camera platforms for research missions that require photographic or video documentation. Aeronautical engineers enable project verification by the widespread use of photos and video.

The F/A-18B support aircraft sends live video back to Armstrong to enable engineers to keep an eye on the mission as it is flown. This photo and video chase ensures and improves flight safety. Armstrong research pilots also use the F/A-18 in the routine flight training of all NASA pilots.

NASA has over the years operated a number of other Hornet aircraft in trials, including many early production and pre-production jets. Most of these have been acquired from the regular forces when they upgraded. These are retired as their airframe life expires and less time expired examples replace them. A further three two-seat Hornets were acquired on 18 November 2018, two coming from VX-23 at Patuxent River and one from VMFA-112 at JRB Fort Worth. Some of these earlier examples have been preserved, while others have been passed into ground training roles. Those known to have been used include:

160780/N840NA
160781/N845NA
160785
161213/N844NA
161214/N817NA & N842NA
161216/N841NA
161217/N852NA
161251
161355/N846NA
161519/N843NA
161520/N847NA
161703/N850NA
161705/N851NA
161744/N835NA & N853NA
161949/N848NA
162414

NASA has operated a number of former Navy F/A-18A/B aircraft in a support role for its ongoing trials. The aircraft have generally been retired by the fleet but still have sufficient airframe hours available. Here the sixteenth production Hornet, 161217, is seen in the NASA colours as N852NA.

NASA has three Hornets assigned to its Edwards AFB facility, a single F/A-18B and a pair of F/A-18A aircraft that it uses for chase purposes. One is 161705, operated as N851NA.

7

CARRIER AIR WINGS

The United States Navy currently operates ten air wings, a significant reduction from the heady days of twenty-one. Thirty to forty years ago the basic air wing was made up of two fighter squadrons, three attack squadrons, one reconnaissance, one electronic warfare, one AEW and one ASW/SAR helicopter squadron, totalling around seventy plus aircraft. This diversity of types must have created a logistics nightmare, with so many to accommodate and maintain.

On top of this, the air reserve component deployed occasionally as an air wing but more often or not supplemented existing wings. Then there was a training carrier, the USS *Lexington*, in the 1970s, but now a normal fleet carrier on work-up is used.

However, the introduction of the Hornet and Super Hornet has simplified this logistics chain and as such created a much leaner and more efficient force structure. Today, the ten air wings are made up of four strike fighter squadrons, one electronic warfare squadron, one AEW squadron and two ASW/SAR helicopter squadrons, comprising three basic types of aircraft, although still maintaining the same number of aircraft.

The VFA-81 commander's aircraft seen in May 1990 just prior to the squadron's first cruise with the Hornet aboard CV-61, the USS *Saratoga*. (Photo Robbie Shaw)

Established at NAS Cecil Field in October 1989, VFA-203 relocated to NAS Atlanta seven years later following the closure of Cecil Field as part of the BRAC programme. This jet, 162868, was acquired from the fleet RAS, VFA-106, and was lost in an accident on 16 January 2001 near Savannah, Georgia.

The current recent structure is as follows:

COMNAVAIRLANT
Commander Naval Air Force US Atlantic Fleet

CVW-1 AB – currently assigned to CVN-75
CVW-3 AC – currently assigned to CVN-78
CVW-6 AE – disestablished 1 April 1992
CVW-7 AG – currently assigned to CVN-72
CVW-8 AJ – currently assigned to CVN-77
CVW-10 AK – disestablished 20 November 1969
CVW-13 AK – disestablished 1 January 1991
CVW-17 AA – transferred to west coast in 2014

COMNAVAIRPAC
Commander Naval Air Force US Pacific Fleet

CVW-2 NE – currently assigned to CVN-70
CVW-5 NF – currently assigned to CVN-76
CVW-9 NG – currently assigned to CVN-74
CVW-10 NM – disestablished 1 June 1988
CVW-11 NH – currently assigned to CVN-68
CVW-14 NK – disestablished 31 March 2017
CVW-15 NL – disestablished 31 March 1995
CVW-16 AH – disestablished 30 June 1971
CVW-17 NA – currently assigned to CVN-71
CVW-19 NM – disestablished 30 June 1977
CVW-21 NP – disestablished 12 December 1975

COMNAVAIRRES
Commander Naval Air Force US Reserves

CVWR-20 AF – Re-described as Tactical Support Wing 1 April 2007
CVWR-30 ND – disestablished 31 December 1994

Left: The twin tails of 164236, belonging to VFA-15. The high-visibility marking is for the CVW-8 CAG on the inboard side of the port fin and the CO of the squadron outboard of both fins.

Below: This F/A-18C 163471 of VFA-83 is seen sporting CVW-17 titles on the outside of the port fin rather than squadron insignia. (Photo via Andy Thomson)

PRODUCTION LIST

McDonnell Douglas/Boeing F/A-18A–D Production – the Bureau Numbers

Bureau Numbers	Designation
160775–160780	F/A-18A-2-MC
160781	F/A-18B-2-MC
160782–160783	F/A-18A-3-MC
160784	F/A-18B-3-MC
160785	F/A-18A-3-MC
161213–161216	F/A-18A-4-MC (Lot 3)
161217	F/A-18B-4-MC (Lot 3)
161248	F/A 18A 4 MC (Lot 3)
161249	F/A-18B-4-MC (Lot 3)
161250–161251	F/A-18A-4-MC (Lot 3)
161353	F/A-18A-5-MC (Lot 4)
161354–161357	F/A-18B-5-MC (Lot 4)
161358–161359	F/A-18A-5-MC (Lot 4)
161360	F/A-18B-6-MC (Lot 4)
161361–161367	F/A-18A-6-MC (Lot 4)
161519	F/A-18A-6-MC (Lot 4)
161520–161528	F/A-18A-7-MC (Lot 4)
161702–161703	F/A-18A-8-MC (Lot 5)
161704	F/A-18B-8-MC (Lot 5)
161705–161706	F/A-18A-8-MC (Lot 5)
161707	F/A-18B-8-MC (Lot 5)
161708–161710	F/A-18A-8-MC (Lot 5)
161711	F/A-18B-8-MC (Lot 5)
161712–161713	F/A-18A-8-MC (Lot 5)
161715	F/A-18A-8-MC (Lot 5)
161716–161718	F/A-18A-9-MC (Lot 5)
161719	F/A-18B-9-MC (Lot 5)
161720–161722	F/A-18A-9-MC (Lot 5)
161723	F/A-18B-9-MC (Lot 5)
161724–161726	F/A-18A-9-MC (Lot 5)
161727	F/A-18B-9-MC (Lot 5)
161728–161732	F/A-18A-9-MC (Lot 5)
161733	F/A-18B-9-MC (Lot 5)
161734–161736	F/A-18A-9-MC (Lot 5)
161737–161739	F/A-18A-10-MC (Lot 5)
161740	F/A-18B-10-MC (Lot 5)
161741–161745	F/A-18A-10-MC (Lot 5)
161746	F/A-18B-10-MC (Lot 5)
161747–161761	F/A-18A-10-MC (Lot 5)
161924	F/A-18B-10-MC (Lot 5)
161925–161931	F/A-18A-11-MC (Lot 6)
161932	F/A-18B-11-MC (Lot 6)
161933–161937	F/A-18A-11-MC (Lot 6)
161938	F/A-18B-11-MC (Lot 6)
161939–161942	F/A-18A-11-MC (Lot 6)
161943	F/A-18B-11-MC (Lot 6)
161944	F/A-18A-11-MC (Lot 6)
161945–161946	F/A-18A-12-MC (Lot 6)
161947	F/A-18B-12-MC (Lot 6)
161948–161965	F/A-18A-12-MC (Lot 6)
161966–161987	F/A-18A-13-MC (Lot 6)
162394–162401	F/A-18A-14-MC (Lot 7)
162402	F/A-18B-14-MC (Lot 7)
162403–162407	F/A-18A-14-MC (Lot 7)
162408	F/A-18B-14-MC (Lot 7)
162409–162412	F/A-18A-14-MC (Lot 7)
162413	F/A-18B-14-MC (Lot 7)
162414	F/A-18A-14-MC (Lot 7)
162415–162418	F/A-18A-15-MC (Lot 7)
162419	F/A-18B-15-MC (Lot 7)
162420–162426	F/A-18A-15-MC (Lot 7)
162427	F/A-18B-15-MC (Lot 7)
162428–162444	F/A-18A-15-MC (Lot 7)
162445–162477	F/A-18A-16-MC (Lot 7)

The first development F/A-18A Hornet was 160775, which was used extensively by the combined trials team at NAS Patuxent River before being 'loaned' to NASA Dryden by April 1985. Retired and struck of charge in 1986, it was delivered to NWC China Lake, where it was subsequently restored and placed on display outside the station's museum.

When the United States Navy celebrated 100 years of naval aviation a number of special schemes were applied across the active fleet. At NAS North Island, the home of the west coast Hornet deep maintenance and repair depot unit, chase plane F/A-18A 162410 received its own commemorative scheme. The jet, once a USMC mount from VMFA-323, was one of the few non-upgraded aircraft still in the inventory.

162826–162835	F/A-18A-17-MC (Lot 8)
162836	F/A-18B-17-MC (Lot 8)
162837–162841	F/A-18A-17-MC (Lot 8)
162842	F/A-18B-17-MC (Lot 8)
162843–162849	F/A-18A-17-MC (Lot 8)
162850	F/A-18B-17-MC (Lot 8)
162851–162852	F/A-18A-17-MC (Lot 8)
162853–162856	F/A-18A-18-MC (Lot 8)
162857	F/A-18B-18-MC (Lot 8)
162858–162863	F/A-18A-18-MC (Lot 8)
162864	F/A-18B-18-MC (Lot 8)
162865–162869	F/A-18A-18-MC (Lot 8)
162870	F/A-18B-18-MC (Lot 8)
162871–162875	F/A-18A-18-MC (Lot 8)
162876	F/A-18B-18-MC (Lot 8)
162877–162884	F/A-18A-18-MC (Lot 8)
162885	F/A-18B-19-MC (Lot 8)
162886–162909	F/A-18A-19-MC (Lot 8)
163092–163103	F/A-18A-20-MC (Lot 9)
163104	F/A-18B-20-MC (Lot 9)
163105–163109	F/A-18A-20-MC (Lot 9)
163110	F/A-18B-20-MC (Lot 9)
163111–163114	F/A-18A-20-MC (Lot 9)
163115	F/A-18B-20-MC (Lot 9)
163116–163118	F/A-18A-20-MC (Lot 9)
163119–163122	F/A-18A-21-MC (Lot 9)
163123	F/A-18B-21-MC (Lot 9)
163124–163145	F/A-18A-21-MC (Lot 9)
163146–163175	F/A-18A-21-MC (Lot 9)
163427–163433	F/A-18C-23-MC (Lot 10)
163434	F/A-18D-23-MC (Lot 10)
163435	F/A-18C-23-MC (Lot 10)
163436	F/A-18D-23-MC (Lot 10)
163437–163440	F/A-18C-23-MC (Lot 10)
163441	F/A-18D-23-MC (Lot 10)
163442–163444	F/A-18C-23-MC (Lot 10)
163445	F/A-18D-23-MC (Lot 10)
163446	F/A-18C-23-MC (Lot 10)
163447	F/A-18D-23-MC (Lot 10)
163448–163451	F/A-18C-23-MC (Lot 10)
163452	F/A-18D-23-MC (Lot 10)
163453	F/A-18C-23-MC (Lot 10)
163454	F/A-18D-23-MC (Lot 10)
163455–163456	F/A-18C-23-MC (Lot 10)
163457	F/A-18D-23-MC (Lot 10)
163458–163459	F/A-18C-24-MC (Lot 10)
163460	F/A-18A-24-MC (Lot 10)
163461–163463	F/A-18C-24-MC (Lot 10)
163464	F/A-18D-24-MC (Lot 10)
163465–163467	F/A-18C-24-MC (Lot 10)
163468	F/A-18D-24-MC (Lot 10)
163469–163471	F/A-18C-24-MC (Lot 10)
163472	F/A-18D-24-MC (Lot 10)
163473	F/A-18C-24-MC (Lot 10)
163474	F/A-18D-24-MC (Lot 10)
163475–163478	F/A-18C-24-MC (Lot 10)
163479	F/A-18D-24-MC (Lot 10)
163480–163481	F/A-18C-24-MC (Lot 10)
163482	F/A-18D-24-MC (Lot 10)
163483–163485	F/A-18C-24-MC (Lot 10)
163486	F/A-18D-24-MC (Lot 10)
163487	F/A-18C-25-MC (Lot 10)
163488	F/A-18D-25-MC (Lot 10)
163489–163491	F/A-18C-25-MC (Lot 10)
163492	F/A-18D-25-MC (Lot 10)
163493–163496	F/A-18C-25-MC (Lot 10)
163497	F/A-18D-25-MC (Lot 10)
163498–163499	F/A-18C-25-MC (Lot 10)
163500–163501	F/A-18D-25-MC (Lot 10)
163502–163506	F/A-18C-25-MC (Lot 10)
163507	F/A-18D-25-MC (Lot 10)
163508–163509	F/A-18C-25-MC (Lot 10)
163510	F/A-18D-25-MC (Lot 10)
163699	F/A-18C-26-MC (Lot 11)
163700	F/A-18D-26-MC (Lot 11)
163701–163706	F/A-18C-26-MC (Lot 11)
163707	F/A-18D-26-MC (Lot 11)
163708–163719	F/A-18C-26-MC (Lot 11)
163720	F/A-18D-26-MC (Lot 11)
163721–163726	F/A-18C-26-MC (Lot 11)
163727–163733	F/A-18C-27-MC (Lot 11)
163734	F/A-18D-27-MC (Lot 11)
163735–163748	F/A-18C-27-MC (Lot 11)
163749	F/A-18D-27-MC (Lot 11)
163750–163754	F/A-18C-27-MC (Lot 11)
163755–163762	F/A-18C-28-MC (Lot 11)
163763	F/A-18D-28-MC (Lot 11)
163764–163770	F/A-18C-28-MC (Lot 11)
163771	F/A-18D-28-MC (Lot 11)
163772–163777	F/A-18C-28-MC (Lot 11)
163778	F/A-18D-28-MC (Lot 11)
163779–163782	F/A-18C-28-MC (Lot 11)
163985	F/A-18C-29-MC (Lot 12)
163986	F/A-18D-29-MC (Lot 12)

163987–163988 F/A-18C-29-MC (Lot 12)
163989 F/A-18D-29-MC (Lot 12)
163990 F/A-18C-29-MC (Lot 12)
163991 F/A-18D-29-MC (Lot 12)
163992–163993 F/A-18C-29-MC (Lot 12)
163994 F/A-18D-29-MC (Lot 12)
163995–163996 F/A-18C-29-MC (Lot 12)
163997 F/A-18D-29-MC (Lot 12)
163998–164000 F/A-18C-29-MC (Lot 12)
164001 F/A-18D-29-MC (Lot 12)
164002–164004 F/A-18C-29-MC (Lot 12)
164005 F/A-18D-29-MC (Lot 12)
164006–164008 F/A-18C-29-MC (Lot 12)
164009 F/A-18D-29-MC (Lot 12)
164010 F/A-18C-29-MC (Lot 12)
164011 F/A-18D-29-MC (Lot 12)
164012–164013 F/A-18C-29-MC (Lot 12)
164014 F/A-18D-29-MC (Lot 12)
164015–164016 F/A-18C-30-MC (Lot 12)
164017 F/A-18D-30-MC (Lot 12)
164018 F/A-18C-30-MC (Lot 12)
164019 F/A-18D-30-MC (Lot 12)
164020–164021 F/A-18C-30-MC (Lot 12)
164022 F/A-18D-30-MC (Lot 12)
164023 F/A-18C-30-MC (Lot 12)
164024 F/A-18D-30-MC (Lot 12)
164025 F/A-18C-30-MC (Lot 12)
164026 F/A-18D-30-MC (Lot 12)
164027 F/A-18C-30-MC (Lot 12)
164028 F/A-18D-30-MC (Lot 12)
164029–164031 F/A-18C-30-MC (Lot 12)
164032 F/A-18D-30-MC (Lot 12)
164033–164034 F/A-18C-30-MC (Lot 12)
164035 F/A-18D-30-MC (Lot 12)
164036–164037 F/A-18C-30-MC (Lot 12)
164038 F/A-18D-30-MC (Lot 12)
164039 F/A-18C-30-MC (Lot 12)
164040 F/A-18D-30-MC (Lot 12)
164041–164042 F/A-18C-31-MC (Lot 12)
164043 F/A-18D-31-MC (Lot 12)
164044–164045 F/A-18C-31-MC (Lot 12)
164046 F/A-18D-31-MC (Lot 12)
164047–164048 F/A-18C-31-MC (Lot 12)
164049 F/A-18D-31-MC (Lot 12)
164050 F/A-18C-31-MC (Lot 12)
164051 F/A-18D-31-MC (Lot 12)
164052 F/A-18C-31-MC (Lot 12)
164053 F/A-18D-31-MC (Lot 12)
164054–164055 F/A-18C-31-MC (Lot 12)
164056 F/A-18D-31-MC (Lot 12)
164057 F/A-18C-31-MC (Lot 12)
164058 F/A-18D-31-MC (Lot 12)
164059–164060 F/A-18C-31-MC (Lot 12)
164061 F/A-18D-31-MC (Lot 12)
164062–164063 F/A-18C-31-MC (Lot 12)
164064 F/A-18D-31-MC (Lot 12)
164065–164067 F/A-18C-31-MC (Lot 12)
164068 F/A-18D-31-MC (Lot 12)
164196 F/A-18D-32-MC (Lot 13)
164197 F/A-18C-32-MC (Lot 13)
164198 F/A-18D-32-MC (Lot 13)
164199–164202 F/A-18C-32-MC (Lot 13)
164203 F/A-18D-32-MC (Lot 13)
164204–164206 F/A-18C-32-MC (Lot 13)
164207 F/A-18D-32-MC (Lot 13)
164208–164210 F/A-18C-32-MC (Lot 13)
164211 F/A-18D-32-MC (Lot 13)
164212–164215 F/A-18C-32-MC (Lot 13)
164216 F/A-18D-32-MC (Lot 13)
164217–164218 F/A-18C-32-MC (Lot 13)
164219 F/A-18D-32-MC (Lot 13)
164220–164223 F/A-18C-33-MC (Lot 13)
164224 F/A-18D-33-MC (Lot 13)
164225–164227 F/A-18C-33-MC (Lot 13)
164228 F/A-18D-33-MC (Lot 13)
164229–164232 F/A-18C-33-MC (Lot 13)
164233 F/A-18D-33-MC (Lot 13)
164234–164236 F/A-18C-33-MC (Lot 13)
164237 F/A-18D-33-MC (Lot 13)
164238–164240 F/A-18C-33-MC (Lot 13)
164241 F/A-18D-33-MC (Lot 13)
164242–164244 F/A-18C-33-MC (Lot 13)
164245 F/A-18D-33-MC (Lot 13)
164246–164248 F/A-18C-33-MC (Lot 13)
164249 F/A-18D-33-MC (Lot 13)
164250–164253 F/A-18C-34-MC (Lot 13)
164254 F/A-18D-34-MC (Lot 13)
164255–164258 F/A-18C-34-MC (Lot 13)
164259 F/A-18D-34-MC (Lot 13)
164260–164262 F/A-18C-34-MC (Lot 13)
164263 F/A-18D-34-MC (Lot 13)
164264–164266 F/A-18C-34-MC (Lot 13)
164267 F/A-18D-34-MC (Lot 13)
164268–164271 F/A-18C-34-MC (Lot 13)

F/A-18C 163443 is seen here in October 2009 as the VFA-125 executive officer's assigned mount sporting a non-standard high-visibility scheme.

164272	F/A-18D-34-MC (Lot 13)
164273–164278	F/A-18C-34-MC (Lot 13)
164279	F/A-18D-34-MC (Lot 13)
164627–164648	F/A-18C-35-MC (Lot 14)
164649–164653	F/A-18D-36-MC (Lot 14)
164654–164655	F/A-18C-36-MC (Lot 14)
164656	F/A-18D-36-MC (Lot 14)
164657–164658	F/A-18C-36-MC (Lot 14)
164659	F/A-18D-36-MC (Lot 14)
164660–164661	F/A-18C-36-MC (Lot 14)
164662	F/A-18D-36-MC (Lot 14)
164663–164664	F/A-18C-36-MC (Lot 14)
164665	F/A-18D-36-MC (Lot 14)
164666	F/A-18C-36-MC (Lot 14)
164667	F/A-18D-36-MC (Lot 14)
164668–164669	F/A-18C-36-MC (Lot 14)
164670	F/A-18D-37-MC (Lot 14)
164671	F/A-18C-37-MC (Lot 14)
164672	F/A-18D-37-MC (Lot 14)
164673	F/A-18C-37-MC (Lot 14)
164674	F/A-18D-37-MC (Lot 14)
164675–164676	F/A-18C-37-MC (Lot 14)
164677	F/A-18D-37-MC (Lot 14)
164678	F/A-18C-37-MC (Lot 14)
164679	F/A-18D-37-MC (Lot 14)
164680–164682	F/A-18C-37-MC (Lot 14)
164683	F/A-18D-37-MC (Lot 14)
164684	F/A-18C-37-MC (Lot 14)
164685	F/A-18D-37-MC (Lot 14)
164686–164687	F/A-18C-37-MC (Lot 14)
164688	F/A-18D-37-MC (Lot 14)
164689	F/A-18C-37-MC (Lot 14)
164690	F/A-18D-37-MC (Lot 14)
164691	F/A-18C-37-MC (Lot 14)
164692	F/A-18D-37-MC (Lot 14)
164693	F/A-18C-38-MC (Lot 15)
164694	F/A-18D-38-MC (Lot 15)
164695–164698	F/A-18C-38-MC (Lot 15)
164699	F/A-18D-38-MC (Lot 15)
164700–164701	F/A-18C-38-MC (Lot 15)
164702	F/A-18D-38-MC (Lot 15)
164703–164704	F/A-18C-38-MC (Lot 15)

164705	F/A-18D-38-MC (Lot 15)
164706–164708	F/A-18C-38-MC (Lot 15)
164709–164710	F/A-18C-39-MC (Lot 15)
164711	F/A-18D-39-MC (Lot 15)
164712–164713	F/A-18C-39-MC (Lot 15)
164714	F/A-18D-39-MC (Lot 15)
164715–164716	F/A-18C-39-MC (Lot 15)
164717	F/A-18D-39-MC (Lot 15)
164718–164722	F/A-18C-39-MC (Lot 15)
164723	F/A-18D-39-MC (Lot 15)
164724	F/A-18C-39-MC (Lot 15)
164725	F/A-18C-40-MC (Lot 15)
164726	F/A-18D-40-MC (Lot 15)
164727–164728	F/A-18C-40-MC (Lot 15)
164729	F/A-18D-40-MC (Lot 15)
164730–164734	F/A-18C-40-MC (Lot 15)
164735	F/A-18D-40-MC (Lot 15)
164736–164737	F/A-18C-40-MC (Lot 15)
164738	F/A-18D-40-MC (Lot 15)
164739–164740	F/A-18C-40-MC (Lot 15)
164865	F/A-18C-41-MC (Lot 16)
164866	F/A-18D-41-MC (Lot 16)
164867	F/A-18C-41-MC (Lot 16)
164868	F/A-18D-41-MC (Lot 16)
164869	F/A-18C-41-MC (Lot 16)
164870	F/A-18D-41-MC (Lot 16)
164871	F/A-18C-41-MC (Lot 16)
164872	F/A-18D-41-MC (Lot 16)
164873	F/A-18C-41-MC (Lot 16)
164874	F/A-18D-41-MC (Lot 16)
164875	F/A-18C-41-MC (Lot 16)
164876	F/A-18D-41-MC (Lot 16)
164877	F/A-18C-41-MC (Lot 16)
164878	F/A-18D-41-MC (Lot 16)
164879	F/A-18C-41-MC (Lot 16)
164880	F/A-18D-41-MC (Lot 16)
164881	F/A-18C-41-MC (Lot 16)
164882	F/A-18D-41-MC (Lot 16)
164883	F/A-18C-41-MC (Lot 16)
164884	F/A-18D-42-MC (Lot 16)
164885	F/A-18C-42-MC (Lot 16)
164886	F/A-18D-42-MC (Lot 16)
164887	F/A-18C-42-MC (Lot 16)
164888	F/A-18D-42-MC (Lot 16)
164889–164897	F/A-18C-42-MC (Lot 16)
164898	F/A-18D-43-MC (Lot 16)
164899–164900	F/A-18C-43-MC (Lot 16)
164901	F/A-18D-43-MC (Lot 16)
164902–164912	F/A-18C-43-MC (Lot 16)
164945	F/A-18D-44-MC (Lot 17)
164946	F/A-18C-44-MC (Lot 17)
164947	F/A-18D-44-MC (Lot 17)
164948	F/A-18C-44-MC (Lot 17)
164949	F/A-18D-44-MC (Lot 17)
164950	F/A-18C-44-MC (Lot 17)
164951	F/A-18D-44-MC (Lot 17)
164952	F/A-18C-44-MC (Lot 17)
164953	F/A-18D-44-MC (Lot 17)
164954	F/A-18C-44-MC (Lot 17)
164955	F/A-18D-44-MC (Lot 17)
164956	F/A-18C-44-MC (Lot 17)
164957	F/A-18D-45-MC (Lot 17)
164958	F/A-18C-45-MC (Lot 17)
164959	F/A-18D-45-MC (Lot 17)
164960	F/A-18C-45-MC (Lot 17)
164961	F/A-18D-45-MC (Lot 17)
164962	F/A-18C-45-MC (Lot 17)
164963	F/A-18D-45-MC (Lot 17)
164964	F/A-18C-45-MC (Lot 17)
164965	F/A-18D-45-MC (Lot 17)
164966	F/A-18C-45-MC (Lot 17)
164967	F/A-18D-45-MC (Lot 17)
164968	F/A-18C-45-MC (Lot 17)
164969–164980	F/A-18C-46-MC (Lot 17)
165171–165182	F/A-18C-47-MC (Lot 18)
165183–165194	F/A-18C-48-MC (Lot 18)
165195–165206	F/A-18C-49-MC (Lot 18)
165207–165218	F/A-18C-50-MC (Lot 19)
165219–165230	F/A-18C-51-MC (Lot 19)
165399–165408	F/A-18C-51-MC (Lot 20)
165409–165416	F/A-18D-51-MC (Lot 20)
165526	F/A-18C-52-MC (Lot 20)
165527–165532	F/A-18D-52-MC (Lot 20)
165680–165687	F/A-18D-52-MC (Lot 21)

UNITED STATES AIRCRAFT CARRIERS

Midway Class

CV-41 USS *Midway*, Commissioned 10 September 1945, Decommissioned 11 April 1992
CV-43 USS *Coral Sea*, Commissioned 1 October 1947, Decommissioned 26 April 1990

Forrestal Class

CV-59 USS *Forrestal*, Commissioned 1 October 1955, Decommissioned 30 September 1993
CV-60 USS *Saratoga*, Commissioned 14 April 1956, Decommissioned 20 August 1994
CV-62 USS *Independence*, Commissioned 10 January 1959, Decommissioned 30 September 1998

Enterprise Class

CVN-65 USS *Enterprise*, Commissioned 25 November 1961, Decommissioned 3 February 2017

Kitty Hawk Class

CV-63 USS *Kitty Hawk*, Commissioned 29 April 1961, Decommissioned 12 May 2009
CV-64 USS *Constellation*, Commissioned 27 October 1961, Decommissioned 8 August 2003
CV-66 USS *America*, Commissioned 23 January 1965, Decommissioned 9 August 1996
CV-67 USS *John F. Kennedy*, Commissioned 8 September 1968, Decommissioned 23 March 2007

Nimitz Class

CVN-68 USS *Nimitz*, Commissioned 3 May 1975
CVN-69 USS *Dwight D. Eisenhower*, Commissioned 18 October 1977
CVN-70 USS *Carl Vinson*, Commissioned 13 March 1982
CVN-71 USS *Theodore Roosevelt*, Commissioned 25 October 1986
CVN-72 USS *Abraham Lincoln*, Commissioned 11 November 1989
CVN-73 USS *George Washington*, Commissioned 4 July 1992
CVN-74 USS *John C. Stennis*, Commissioned 9 December 1995
CVN-75 USS *Harry S. Truman*, Commissioned 25 July 1998
CVN-76 USS *Ronald Reagan*, Commissioned 12 July 2003
CVN-77 USS *George H.W. Bush*, Commissioned 10 January 2009

Gerald R. Ford Class

CVN-78 USS *Gerald R. Ford*, Commissioned 22 July 2017
CVN-79 USS *John F. Kennedy*, due 2022
CVN-80 USS *Enterprise*, construction to start 2018, due 2027

CV-41 the USS *Midway* served for many years as the Yokohama-based Pacific fleet carrier. Retired from active duty, the ship has been preserved in San Diego as a permanent museum dedicated to everything US Navy.

The USS *Dwight D. Eisenhower* seen in Stokes Bay in the Solent, United Kingdom.

CVN-70, USS *Carl Vinson* seen at rest in San Diego harbour. Note the legacy Hornet on deck in use as 'spotting dolly' to train aircraft handlers.

A look across the bow of CVN-71, the USS *Theodore Roosevelt*, as it heads into the sunrise while in the eastern Mediterranean during Operation Iraqi Freedom.

The USS *Theodore Roosevelt* in Stokes Bay in 2011.

The USS *Ronald Reagan* in San Diego harbour in 2014.

CVN-77 the USS *George H.W. Bush* in Stokes Bay in 2011.

CVN-77 the USS *George H.W. Bush* in Stokes Bay in 2011.

ABBREVIATIONS

ATARS	Advanced Tactical Airborne Reconnaissance System
BRAC	Base Realignment and Closure
CAG	Commander Air Group
CO	Commanding Officer
COMPTUEX	Composite Training Unit Exercise
CV	Aircraft Carrier
CVA	Aircraft Carrier Attack
CVN	Aircraft Carrier Nuclear
CVW	Aircraft Carrier Wing
EastLant	Eastern Atlantic
EO	Electro Optical
FLEETEX	Fleet Exercise
FRS	Fleet Replacement Squadron
GD	General Dynamics
HARM	High-speed Anti-Radiation Missile
JDAM	Joint Direct Attack Munition
JTAC	Joint Terminal Attack Controller
JTFEX	Joint Fleet Exercise
LGB	Laser-Guided Bomb
LEX	Leading Edge Extension
MAG	Marine Air Group
MCAS	Marine Corps Air Station
MFD	Multi-Function Display
NAD	Naval Air Depot
NAF	Naval Air Facility
NAS	Naval Air Station
NASA	National Aeronautics and Space Administration
NATC	Naval Air Test Centre
NAWC	Naval Air Warfare Centre
NorLant	North Atlantic
NorPac	Northern Pacific
NSAWC	Naval Strike and Air Warfare Centre
RAG	Replacement Air Group
RAS	Replacement Air Squadron
RDT&E	Research Development Test & Evaluation
RIMPAC	Pacific Rim
SFARP	Strike Fighter Advanced Readiness Program
SOC	Struck off charge
SUSTEX	Sustainment Exercise
UDP	Unit Deployment Program
USMC	United States Marine Corps
USN	United States Navy
USS	United States Ship
VF	Fighter Squadron
VFA	Fighter Attack Squadron
VFA(N)	Fighter Attack Squadron (Night)
VFC	Fleet Fighter Squadron Composite
VMFA	Marine Fighter Attack Squadron
VMFA(AW)	Marine Fighter Attack Squadron (All Weather)
VX	Fleet Experimental Test Squadron
WestPac	Western Pacific
WTI	Weapons Training Instruction course

SOURCES

Reference Work

Jane's all the World's Aircraft 1988–1989
International Air Power Review, vol. 2, AirTime Publishing 2001
World Air Power Journal, vol. 1 through vol. 43 Aerospace Publishing 1990–2000
Stars and Stripes newspaper, April 2017
Sea Power magazine, various issues
F/A-18 Hornet by Roy Braybrook, Osprey 1991
F-18 Hornet by Herman J. Sixma & Theo W. Van Geffen, Ian Allan 1993
United States Air Forces Directory, 2008 through 2017, Mach III
British Aviation Review, 1976 through 1993 British Aviation Research Group
Tip of the Spear: U.S. Navy Carrier Units and Operations 1974-2000 by Rick Morgan, Schiffer, 2007

www.joebaugher.com
www.gonavy.jp
www.public.navy.mil/airfor/vfa15/Pages/default.aspx
(this applies to most US Navy units as applicable)
www.3rdmaw.marines.mil/Units/MAG-13/VMFA-122/VMFA-122-History/VMFA-122
(an example of unit websites)
www.mag31.marines.mil/MAG-31-units/VMFA-251/About/VMFA-251
(an example of unit websites)
https://www.nasa.gov/centers/armstrong/news/FactSheets/FS-006-DFRC.html
www.mybaseguide.com/navy/91-753-5632/nas_oceana_shore_units

Many of these unit web pages have changed over time as they are updated with some information being deleted. This has certainly been the case with units upgrading from the legacy series aircraft.

Other information has been gathered from base publications and press releases, along with notes taken at the time on visits to various squadrons, aircraft carriers and bases.

Sepecat Jaguar: Almost Extinct

PETER FOSTER

The Jaguar was an iconic aircraft to come from Anglo-French collaboration and one of the first to be conceived with a predatory attack and low-level strike capability. First planned as a trainer, it emerged as a fighter bomber taking much from the TSR2 concept when a string of cancelled projects identified a gap in strike/attack capability; it soon evolved into a supersonic aircraft ready for reconnaissance and tactical nuclear strike roles. Retired before its time, for France in 2005 and for the RAF in 2007, it is still revered both by those that operated it and those that stared in wonder. The end for the Jaguar in the United Kingdom was sudden and rushed with the big cat going out with a meow rather than a roar. However, it survived on other continents providing a growl and bite in maintaining sovereignty for several decades on. This book is a stunning pictorial tribute to those final days.

978 0 7509 7021 1